D0191817

The Bombe at Bletchley

On 4 September 1939 my uncle, Alan Turing, reported for duty at the Government Code & Cypher School at Bletchley Park. He is rightly celebrated as having pushed the frontiers of computing, artificial intelligence and morphogenesis. But it is probably for his work at Bletchley Park, where the British succeeded in breaking into the secret enciphered messages sent between the Axis forces, that he is best known.

In 1939 the immediate challenge was to find a way to read messages enciphered on the Enigma machine. The recruiters for GC&CS recognised that new skills were needed to tackle mechanical encryption: mathematicians were hired as well as linguists. In particular, they hired mathematicians with a vision of machinery which could go quickly and reliably through thousands of permutations too tedious for humans to handle. Together with another mathematician, Gordon Welchman, Alan Turing designed a new type of machine, the Bombe, which would give the codebreakers a breach in the security wall constructed by the Enigma.

The Bombe was not a computer. But Bletchley Park did pioneer the mechanisation of problem-solving. After the Bombe, the codebreakers developed the Colossus, a new breed of electronic machine for breaking into the Lorenz ciphers. Bletchley Park thus became the birthplace of British post-war computing. This is a story in which my uncle played a continuing role, building and using some of the world's first stored-program computers.

This book explains – for visitors who can see the replica Bombe in operation at Bletchley Park today – the basics of the Bombe's design and how it helped with the cracking of Enigma messages. It is a small tribute, but a great privilege to be able to bring this one aspect of the work of a great man to a wider audience.

Sir John Dermot Turing
Bletchley Park, April 2014

Demystifying the Bombe

This book explains to a general reader the challenges of the Enigma machine and how the Bombe was structured to tackle them.

The main purpose is to get an understanding of how the Bombe did its job. It was an astonishing achievement for the Bletchley Park codebreakers to invent, build and put to use a machine which completely undermined the security of the German Enigma ciphers. Given the complexity of the Enigma machine itself, one might be tempted to assume that understanding the machine which unravelled its secrets is a task too difficult to tackle. Actually, the Bombe is quite straightforward in what it did:

- It found the relative starting position of the Enigma's rotors by trying out each possible position in turn and testing it.
- It found one of the cross-pluggings of the Enigma's plugboard.
- It confirmed the rotor selection and the order of rotors.

And there's a lot that the Bombe didn't actually do:

- solve the rest of the plugboard, or the ring-settings;
- convert enciphered intercepts into plain German decrypts.

These things had to be done by the codebreakers using other manual or mechanical techniques. But what the Bombe had done was to open the door so that those techniques could come into play, and thus provide the daily Enigma settings with astonishing speed.

This book explains how the Bombe worked and how its results fitted into the overall process for finding the daily key for a given network of Enigma messages.

The main focus is on the German army and air force Enigma, which yielded a fabulous crop of intelligence.

Front view of a Bombe during the war.

G.C.H.Q.
DECLASS AGREED
26 NOV 2005
HISTORIAN

D1486-001 CCR 1
102(1)

TOP SECRET
ULTRA

65/4/7A

I. A description of the machine.

We begin by describing the 'unsteckered enigma'. The machine consists of a box with 26 keys labelled with the letters of the alphabet and 26 bulbs which shine through stencils on which letters are marked. It also contains wheels whose function will be described later on. When a key is depressed the wheels are made to move in a certain way and a current flows through the wheels to one of the bulbs. ~~Txxxxxxxxxxxxxxx~~ The letter which appears over the bulb is ~~xxiixx~~ the result of enciphering the letter on the depressed key with the wheels in the position they have when the bulb lights.

To understand the working of the machine it is best to separate in our minds

The electric circuit of the machine without the wheels.

The circuit through the wheels.

The mechanism for turning the wheels and for describing the positions of the wheels.

The circuit of the machine without the wheels.

Fig 1

Eintritts walz

The machine contains a cylinder called the Eintrittswalz (E.W) on which are 26 contacts $C_1, \ldots, C_{26}$. The effect of the wheels is to connect these contacts up in pairs, the actual pairings of course depending on the positions of the wheels. On the other side the contacts $C_1, C_2, \ldots, C_{26}$ are connected each to one of the keys. For the moment we will suppose that the order is ~~xxwrtzxixxxdfghjk~~ QWERTZUIOASDFGHJKPYXCVBNML , and we will say that Q is the letter associated with C_1 , W that associated with C_2 etc. This series of letters associated with $C_1, C_2, \ldots, C_{26}$ is called the diagonal, for reasons which will appear in Chap

How this book is arranged

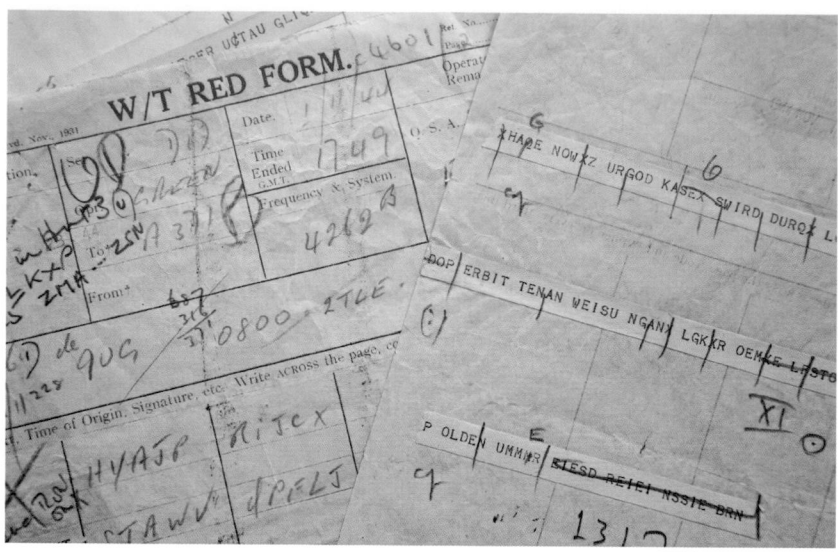

An intercepted message and the raw output of the decryption process.

From puzzle

The Enigma machine posed various challenges. Some were amenable to attack with a Bombe. These were the starting positions of the three rotors which transmuted plain-text into a different letter, revolving with every keystroke to generate a new substitution, and the plugboard, which carried out additional swap-overs of letters.

See pages 6–9

Finally there were the remaining settings of the Enigma to resolve. These were the ring-settings, which determined where the rotor's wiring lay relative to the letters around its rim, and when the middle and left-hand rotors moved on to the next position. This problem could also be solved using the checking machine.

See pages 48–51

Once the settings were known, intercepted messages could then be decrypted using modified 'Typex' machines, and put to use as intelligence. Several tactical successes can be directly linked to Enigma-based intelligence – as well as some missed opportunities.

See pages 52–58

The German navy – significantly the U-boat arm – used Enigma differently. Naval Enigma was, in consequence, much harder to break. Additional techniques, derring-do, and ultimately special 4-rotor Bombes were needed to win the U-boat code-breaking war.

See pages 59–61

The three rotors were the smaller challenge. There are only 17,576 ways of configuring the rotors, if you already know which rotors are in use and the order in which they have been inserted into the Enigma machine. The Bombe tackled this problem simply by going through all 17,576 possibilities one by one – this only took about 10 minutes for a complete run, assuming no stops.

See pages 10–26

The plugboard was a much tougher challenge, with 150 million million solutions. This couldn't be done with exhaustive (brute-force) analysis of all the possibilities. The Bombe tested just one possibility: that a given letter in the crib – the codebreakers' guess at what the intercepted cipher-text actually stood for – was cross-plugged to one letter selected by the codebreakers. The Bombe verified this, or suggested an alternative plug-mate.

See pages 27–42

That left the remaining plugboard connections to be found. A supplementary machine, called the checking machine, could be used to help with the task. At the same time the checking machine confirmed that the rotor configuration and plugboard connection suggested by the Bombe was not a coincidental (but false) solution.

See pages 43–47

After the war, the secrets of Bletchley Park had to be preserved. The original Bombes were disposed of. A replica Bombe was painstakingly reconstructed from previously secret sources.

See pages 62–63

To intelligence

Enigma intelligence was disseminated in the field via Special Communications Units, such as this Packard sedan adapted as a mobile wireless station, operating in Libya near the headquarters of the British Eighth Army.

Know your Enigma

Operating the Enigma is incredibly simple: you press a key on the keyboard, and a different letter lights up in the panel just behind it. The illuminated letter is the enciphered text. To decipher the message, it's exactly the same procedure: using the same Enigma set-up, just press the letter of the cipher-text, and the deciphered text lights up in the panel.

Taking a look under the cover of the Enigma machine, there are four essential components which turn the plain-text into apparently random nonsense.

1   *The set of rotors*. Enigma machines used by the German army and air force had three of these rotors. Each of them has 26 input connectors and 26 output connectors, one for each letter of the alphabet. The rotor has internal wiring so that an electric current coming in at any given letter, say 'N', will be re-routed, coming out at, say, 'D'. This output goes on to become the input for the next of the three rotors, again coming out at any letter. At the end of the third rotor is a fixed reflector plate, which also has internal wiring, sending the current back through the three rotors in the opposite direction, and again potentially changing the letter at each step.

Each rotor is wired differently, and the rotors are interchangeable. So the Germans could vary which rotor they used in which slot in the machine; they had a library of five rotors to choose from, giving 60 possible permutations for choosing the three rotors to place in the machine.

As you might expect, the rotors rotate. Each press of the keyboard key moves the right-hand rotor on one place, so if the operator pressed a second key, the current would enter the rotor system at a different

A 3-rotor Enigma machine with the inner lid lifted and front flap down, to show the rotors, lampboard, keyboard and plugboard.

Cross-pluggings on the plugboard.

Army and air force rotors had numbers around their rims.

place and follow a completely different path, giving a different read-out on the illuminated panel. What's more, each rotor has a notch in its rim which, when reaching a particular position in its rotation, moves the next rotor along one place – what the codebreakers called a 'turnover'; an analogy is how the 'tens' on a counter move up one when the 'units' switch from 9 to 0. So the wiring doesn't just change every letter, but to avoid repetition it undergoes a further reshuffle some time during every run of 26 characters, and also every 676 characters. With a 3-rotor Enigma machine there are 17,576 (26^3) possible wirings based on the rotor configurations alone.

The set of rotors make up the central scrambler – the heart of the Enigma.

Each rotor was cross-wired internally to scramble the message.

Rotors

After a rotor has been inserted into the Enigma machine, it has to be moved round so that the starting position is the same for both sender and receiver of the message. Each rotor has 26 possible starting positions, visible through the small windows in the inner lid of the machine.

Various different rotor types were used by the German armed forces. Rotors labelled with numbers for each of the 26 start positions and ring-setting were typically used by the army and air force, whereas the navy used rotors labelled with letters.

The codebreakers didn't call them 'rotors': their documents refer to 'wheels' and 'wheel orders'. This book follows the more modern usage.

2   *The ring-setting.* The position of the turnover notch could be set to any of 26 places on any of the rotors. This was done by means of a ring attached to the rotor, which could be unclipped and turned round to adjust the turnover points for a particular message, or a particular group of messages. In practice the ring-setting was changed only once every 24 hours.

The sender chose the starting position of the rotors before enciphering the message. To make sure the recipient of a message had his machine set up the same way as the sender, it was also necessary for the sender to tell the recipient the starting positions he had chosen. These can be changed by rotating the finger-grips on the rotors which protrude through the protective inner casing of the machine, until the correct three letters show in the viewing windows.

The ring-setting also had another effect, which was to offset the actual starting position relative to the core of internal wiring in the rotor. That meant that even if we knew the letters showing up in the viewing windows, we would not know how the rotor cores started out at the beginning of the message, unless the ring-settings were known as well. In order to decipher the messages of the day, the codebreakers would have to know the starting position chosen by the Engima operator and the day's ring-settings.

As there are 26 different locations for the turnover notch on each rotor, changing the ring-settings gives rise to another 17,576 set-ups for the Enigma machine, to confound the codebreakers further.

3 *The plugboard.* This was the truly fiendish part of the machine. 17,576 sounds like a lot of permutations, but a machine can easily, if unimaginatively, clunk through all of them given enough time and fast enough technology. The Bletchley Bombes did exactly that: try out all 17,576 rotor configurations systematically in what these days

is called a brute-force attack. So to give the Enigma more security the Germans added a plugboard at the front of the machine. The plugboard connected pairs of letters to each other and had the effect of switching them, one for the other. The plugboard was wired up to the input/output terminals of the scrambler unit, so the current passed through the plugboard twice on its way between the keyboard and the illuminated lampboard.

The plugboard settings were only changed once a day, so they remained constant for the whole collection of a day's messages sent on a given network. For much of the war, the Germans used 10 plugboard connections, which multiplied the number of ways the Enigma could be set up by 150 million million times ($26!/10!6!2^{10}$). The plugboard was going to need more than brute force to crack it.

As the rotor moved round, the notch would eventually engage with a lever meaning the adjacent rotor turned round one place too. 'Turnovers' caused difficulties for the codebreakers, in particular making it harder to design menus.

Crib, Bombe and checking machine

So, to decipher messages, the codebreakers needed to be able to find out: which rotors were being used in the machine, and in which positions; the ring-settings; the plugboard connections and the starting position of the rotors chosen by the operator. In total that was 158.9 million million million[1] false set-ups to eliminate.

The codebreakers at Bletchley Park used a variety of different techniques to tackle this seemingly impossible problem.

The first step was to find a *crib* – a likely piece of plain-text which corresponds to a string of characters in the intercepted message. The Germans were very helpful to the cribsters at Bletchley Park, because many messages contained stereotyped greetings or standardised content. The Enigma machine has no number keys, so numbers had to be typed out in full.

---

1. 60 rotor and slot choices * 17,576 ring-settings * 26!/10!6!2^{10} plugboard choices.

Students learn how the trios of drums on the front of the Bombe replicate the behaviour of the rotors in an Enigma machine.

A rebuilt checking machine in the museum at Bletchley Park.

Using the crib, the Bombe would be connected up and then crank through the 17,576 possible rotor configurations. When the Bombe stopped in the middle of its run, it meant that the Bombe had found a plausible start-up configuration for the three rotors, together with a single possible plugboard connection.

Finally, a smaller machine, called the checking machine, would be used to see if a result produced by the Bombe – a *stop* – was correct or just a chance stop, and also to find out the remaining plugboard settings.

Each part of the Bombe had a vital role in breaking into the daily settings of the Enigma machine. The most plainly visible features are:

- the drums, which mimic the rotors in the Enigma's scrambler unit, and rotate through each of the possible 17,576 configurations, so that they are all tested exhaustively until a contender for the right one is found;
- the cables at the back, which connect together the mock Enigmas constituted by the trios of drums.

The mock Enigmas were connected in accordance with a diagram called a *menu*. The menu was formed from the codebreakers' guess at the plain-text that might have lain behind the intercepted message: if their guess was right, then the Bombe would stop, revealing a plausible starting position of the Enigma rotors and one of the plugboard connections.

Roll of the drums

The Bombe's most striking feature is the three banks of replica rotors on its front. You can see 36 sets of three drums. Each drum contains a replica of the internal wiring of an Enigma rotor, so each set of three drums together replicates the operation of the scrambler in an Enigma machine. During the war, these groups of three drums were called *Letchworth Enigmas*, after the town where the Bombes were manufactured.

The Bombe mechanically rattles through all 17,576 possible configurations of the three rotors in order to find a logically plausible set-up. When the Bombe is running, you can see the drums rotating through these configurations until it stops, which indicates that the Bombe has found a plausible solution – called a 'story' at Bletchley Park. You can hear the Bombe move the middle set of drums about every second; in each

second the fast drum has cycled through all 26 letters, and that means that the Bombe can test all 17,576 configurations in about 10½ minutes. The speed of the Bombe – using electromechanical technology, rather than the modern electronics used to crack the more sophisticated Lorenz cipher in later developments at Bletchley Park – was fast enough to facilitate a brute-force attack to disclose the relative configuration of the wiring in the three rotors in use for the day.

There is also an additional set of three drums at the right-hand end of the middle bank. These are the indicator drums: when the Bombe stops, these show where the starting positions of the Enigma rotors would have been, relative to the starting positions of the drums on the Bombe.

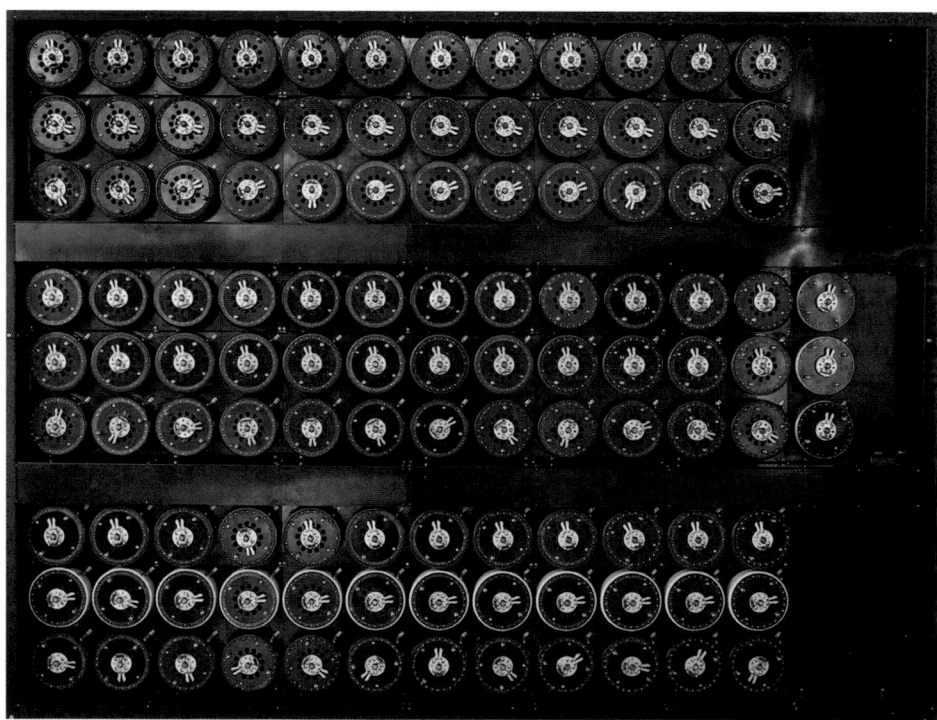

The front of the Bombe. Different combinations of drums could be used in each row to test for different rotor combinations with the same menu.

The reverse of a drum, showing its precise electrical connections. Wrens used eyebrow tweezers to straighten the wires so as to avoid short-circuits.

There was no guarantee that the 'story' offered by the Bombe was correct, and even if it was right there were other parts of the set-up that needed to be solved. But it was a start, and assuming the story provided by the Bombe was right, it was like a breach in a dam: one crack and the whole edifice can break up.

To get the drums rolling for codebreaking, you need to know when you've got a good – or at least plausible – starting position for the rotors. Identifying a plausible configuration is done using a menu derived from an intercepted message and its crib.

The front face of the Bombe without its drums, showing the fixed electrical contacts.

'Bombe'? Why on earth is it called that?

The French chef Auguste Escoffier (1846–1935) was not a codebreaker, but everyone at Bletchley Park would have heard of this celebrity figure. His *Guide Culinaire* of 1907 gives over 60 recipes for *bombes*. These are confections of sponge, ice-cream and meringue, frequently frozen into a mould which is spherical and finished off with a layer of chocolate so as to resemble the classic bomb with a sputtering fuse – an image which everyone has in their memory. A *bombe surprise* is a bombe that arrives at the table from the oven, or even *flambéed* at the table, without the ice-cream having melted.

All right, but what's that got to do with Enigma machines? We don't know. We do know that the Poles, who invented the Bletchley Bombe's predecessor, christened their machine for breaking into message settings a *bomba*, which is the Polish translation of *bombe* – and that could mean either the explosive variety or the edible one in Polish too. The Poles were collaborating effectively with the French

throughout the 1930s, and they may have shared recipes as well as codebreaking know-how: it was customary at the time for Poles to go out and eat just a dessert – such as a scoop (in Polish, a *bomba*) of ice-cream. We also know that '*bomba!*' is what you exclaim in Polish when you have got something which is really, really good: like when Frenchmen say '*super!*', Germans '*prima!*', or Brits 'I say, that's not bad, not bad at all'.

But while we believe the British called their machine a 'Bombe' in tribute to the Poles, that still doesn't explain why the Polish codebreakers called their machine a *bomba*. Maybe it was because the *bomba* ticked away menacingly as it went through its revolutions; maybe the inspirational thoughts which led to its design came along with a good scoop of ice-cream. Or maybe it was just that the Polish codebreakers' cookery was a first-rate scoop.

It's another one of the mysteries connected to Bletchley Park.

Cribbing

The codebreakers entered the secret world of the German Enigma messages by guessing at their content, often relying on stereotyped phrases.

A crib was a piece of text believed to be the plain-text corresponding to part of the intercepted message. Without cribs, the Bombe would not have worked: the fundamental principle of the Bombe is to test whether there was a starting position for the rotors of an Enigma machine which would consistently encipher several letters in the same message into the intercepted cipher-text.

Cribs could be matched to intercepts by exploiting one weakness in the Enigma's design: the fact that no letter can be enciphered as itself.

Hut 6, where the cribsters worked on army and air force Enigma messages, as it was in 2012 before restoration began.

Message position	1	2	3	4	5	6	7	8	9	10	11	12	13	14	15	16	17	18	19	20
Intercepted cipher-text	S	L	Y	A	R	O	P	X	J	Q	K	E	B	K	Q	Z	A	N	Y	G
Crib (possible plain-text)	K	E	I	N	E	X	B	E	S	O	N	D	E	R	E	N	X	E	R	E

In the example above it is assumed that the sender wanted to transmit the rather dull phrase '*Keine besonderen Ereignisse*' (no unusual goings-on), with the letter 'X' being used as a spacer. The first step is to match up the intercept with the crib. Because the Enigma machine does not allow any letter to be enciphered as itself, the crib will only be a possible solution if there is no match between the intercept and the guessed-at crib.

The codebreakers chose a run of letters from a possible crib. For each rotor configuration the Bombe would test whether it was possible for the 3-rotor scrambler unit to convert all the chosen letters from the crib into the intercepted cipher-text at their respective positions in the message.

If the crib is a good guess, then, looking at the positions coloured yellow in the example:

'X' is enciphered to become 'O' when the rotors had moved round 6 places from the start;
'O' is enciphered to become 'Q' 10 places from the start;
'E' is enciphered to become 'Q' 15 places from the start; and
'E' is enciphered to become 'X' 8 places from the start.

These four encipherments can be joined up in the form of a loop, as shown on page 15.

Loop connecting encipherments

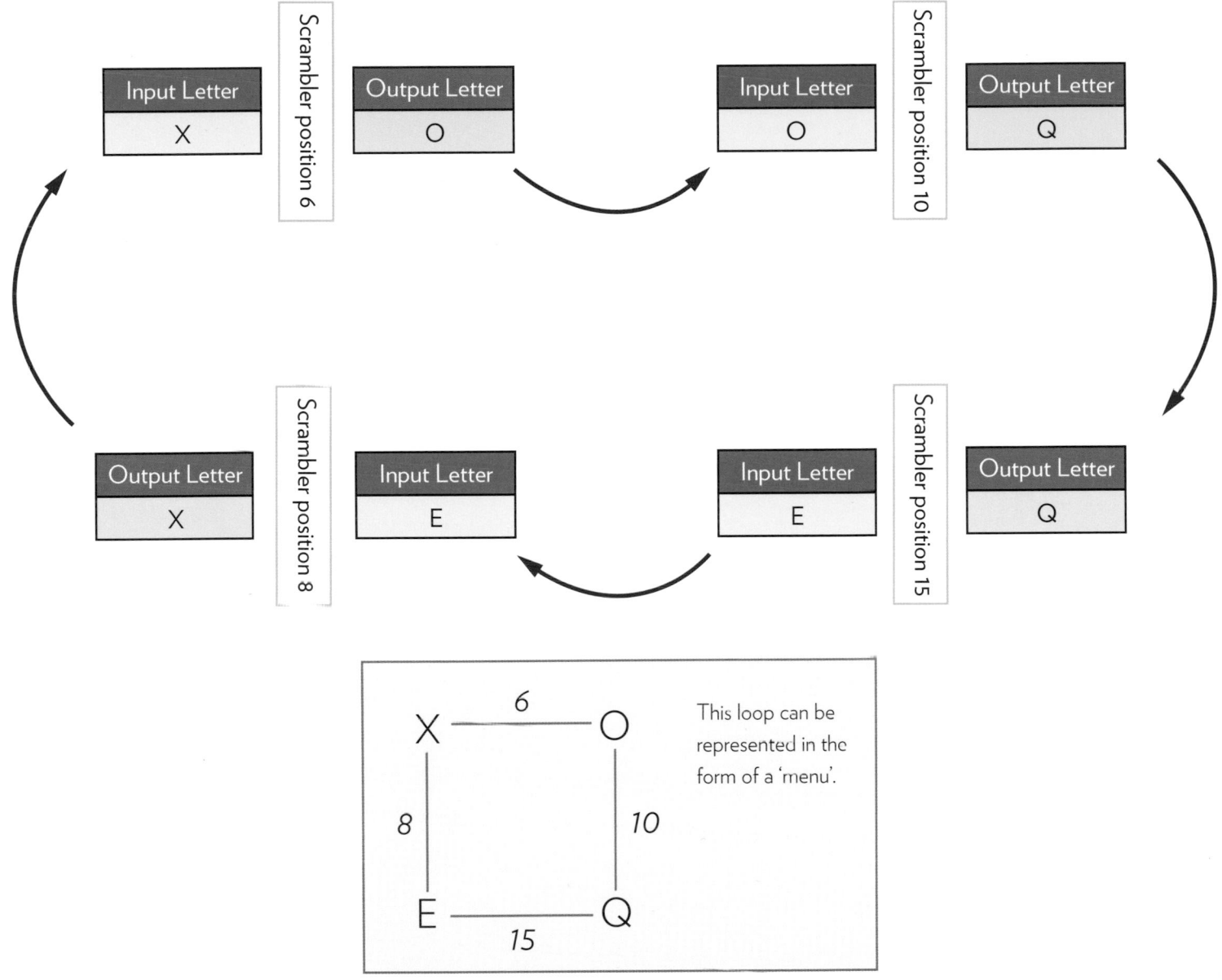

Input Letter
X

Scrambler position 6

Output Letter
O

Input Letter
O

Scrambler position 10

Output Letter
Q

Output Letter
X

Scrambler position 8

Input Letter
E

Input Letter
E

Scrambler position 15

Output Letter
Q

This loop can be represented in the form of a 'menu'.

Cheating at exams

The Shorter Oxford Dictionary records uses of the word 'crib' going back to 1597. As well as meaning 'a small bed for a child', it has some other exotic interpretations, including 'a barred receptacle for fodder', 'a wickerwork basket, pannier or the like', and 'an apparatus like a hay-rack for draining the salt after boiling'. The codebreakers of Bletchley Park did not have any of those obscure senses in mind. The *Oxford Dictionary* puts their usage rather sniffily at the end of its list:

16. Something cribbed; a plagiarism (colloq.) 1834.
17. A translation of a classic, etc., for the illegitimate use of students (colloq.) 1827.

The codebreakers might have doubted whether many of the intercepted messages were classics, but many of them knew of the practice of taking translations of Latin poetry into exams to assist with the process of deciphering unintelligible texts.

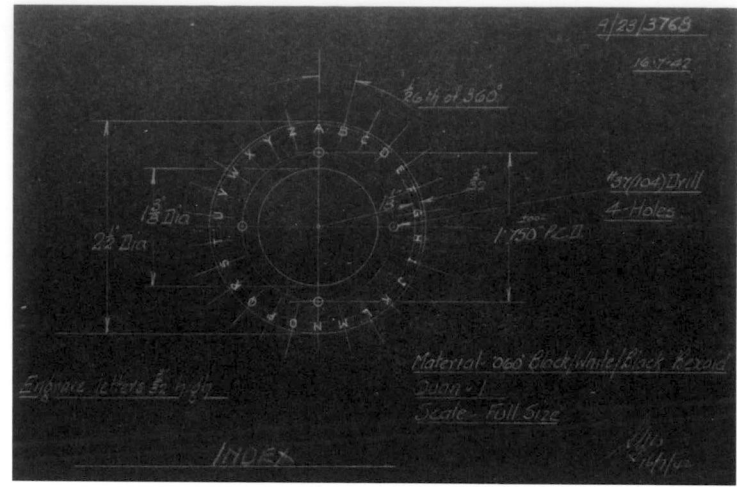

A 1942 Bombe blueprint giving detail for the front plate of an indicator drum. © Crown Copyright – Used with permission of Director GCHQ.

At position 15, the loop is joined up using the 'reciprocity principle': that when the Enigma machine is set up for enciphering 'E' to become 'Q', exactly the same Enigma set-up must enable decipherment of 'Q' into 'E'.

Loops such as this are used to create menus, which provided maps to enable the Wrens who operated the Bombes to connect up the sets of drums – the Letchworth Enigmas. The input and output letters for each set represent a pair of letters – the plain-text and its corresponding cipher-text – as selected from the message. Once connected, the drums could test simultaneously all the chosen paired enciphements, to see if they could all be consistently enciphered on an Enigma machine set up with its rotors in the same starting configuration.

A Bletchley Park volunteer shows how to match up an intercepted message with a possible crib.

The consistency test is part of it; but making a menu from loops of paired enciphements hints at another key design principle of the Bombe: to create an *electrical loop*, or a circuit. In this example, four separate drum-trios on the Bombe can be adjusted to test the same starting point for the message, by moving them on round to correspond to positions 6, 8, 10 and 15 of the intercepted message.

Testing for the rotor configuration

If the drums are in the right configuration for the start of the message, then we would know what their configuration ought to be at each subsequent position in the message. So, taking the example in the table on page 14:

- An Enigma machine set up the right way would encipher the first letter of the plain-text message 'K' as 'S', in the starting position. On the next keystroke the Enigma will move the right-hand rotor on one place, and encipher 'E' into 'L'.

- We could manually move the right-hand rotor of the same Enigma onto another letter in the message, say the fourth one, where 'N' would be enciphered as 'A'; but instead of manually moving the rotors one at a time on a single Enigma machine, we can set up a bank of Enigmas – or rather, Letchworth Enigmas taking the form of Bombe drums – each of which is moved on round from the starting point of the message by the requisite number of positions.

Original cabling and drum components from a wartime Bombe and checking machine.

The interior of the Bombe gives an idea of the complexity of its engineering.

As the Bombe cranks through all 17,576 possible starting positions, the relative orientation of the four sets of drums will stay the same.

Now the Bombe operator can connect up four Letchworth Enigmas with electrical cables in a loop, so that all four can be tested simultaneously. If the guess about the true meaning of the intercepted message is correct, then electricity will flow around the four mock-up Enigmas in a perfect circuit when the drums are in the correct configuration – in other words, when the correct starting positions of the rotors have been found. If the configuration of the drums is wrong, the chances are that at any one set of drums the output will not be correct, and that will create a false input for the next set of drums, and so on; by the time electricity reaches the final connecting cable some letter other than 'X' will have become electrically live. And that will make another electrical path through the drums live.

Hum drum facts

The Bombe drums aren't exactly the same as Enigma rotors. Each drum has two identical lots of internal wiring, one for the inward flow of current through the scrambler, and another for the outward flow; in the Enigma scrambler, the reflector plate ensured a different path back through a single set of wiring in the rotors. Incidentally, in this property of the reflector lies one of the famous weaknesses of the Enigma: that no letter can be enciphered as itself. The role of the reflector in the Bombe is carried out in the rather nondescript boxes that bulge out of the machine at the opposite end to the control panel.

Another difference: Bombe drums don't have ring-settings. Changing the ring-setting offsets the core internal wiring of the rotor relative to the letters round the rim. On the Bombe, each drum is wired to behave like a rotor with the ring-setting at the letter 'Z'. So the starting configuration of rotors suggested by a 'story' when the Bombe stops isn't the actual configuration: it's the configuration relative to the notional ring-setting ZZZ used for the three drums. This setting can be used to decipher one message – the one which provided the crib.

Betting in cycles

The Bombe's test for consistency between crib and intercept relies on connecting the drums together in electrical loops.

- If the starting configuration of rotors is good, then electric current will flow round the loop and come out of the final drum-set in the loop at the same place as it started and no other. This neat state of affairs causes the Bombe to stop.
- If the starting configuration is bad, the current will come out somewhere else, and because the sets of drums are connected up in a loop, the live output wire at the 'end' of the loop will feed electricity into a new input terminal on the first set of drums, leading to a messy situation where other wires in the cabling become live. So if all 26 wires are live, the Bombe knows that the configuration is no good, and moves on to test the next rotor starting configuration.

The test works like an accumulator in horse-racing. Using the example in the table, let us select the four encipherments coloured yellow and set up four Letchworth Enigmas, one for each position in the message, duly offset from the starting point by the number of positions corresponding to its place in the message. The output from Letchworth Enigma is used as the input to the next. In this particular example, the output from the drums offset at position 6 is used as input to the drums offset at position 10; the output from the drums offset at position 10 can be used as input to the drums offset at position 15, and so on.

If the crib is a good guess at the actual plain-text, when the first rotor is in the correct alignment an input at letter 'X' in the first Letchworth Enigma (set to position 6) will come out at letter 'O' and the current will then go into the second Letchworth Enigma (set to position 10)

Many Bombes were actually located in outstations. This is Hut 11A, where some of the Bombes were located within the Bletchley Park site.

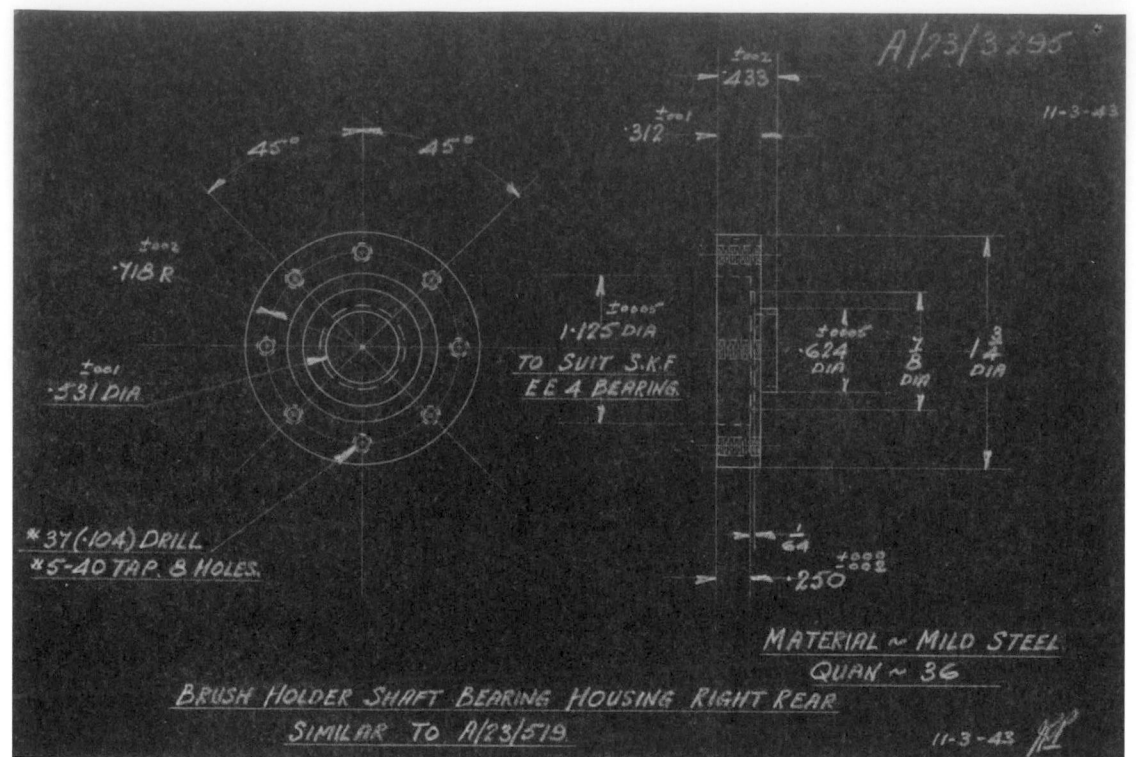

Brush Holder Shaft Bearing Housing Right Rear
Similar to A/23/519

A blueprint from 1943 indicates the care in design needed to achieve electrical precision at high speed. © Crown Copyright – Used with permission of Director GCHQ.

at letter 'O'. But only if this second set of drums is also in the correct alignment will the current come out at letter 'Q'. If that happens, our accumulator is working, and if all four Letchworth Enigmas are correctly configured, the current will come out at letter 'X' from the fourth set of drums.

In any other alignment the input at letter 'X' will come out somewhere else, and we've lost our bet, so the Bombe will move on to the next possible alignment.

The idea of the loops – having the input current cycle round and come out in the wrong place if the drum configuration was bad – was a central feature of the Bombe's design.

Hut 6, where army and air force menus were designed, after restoration in 2014.

The feedback loop

If the drums are in the right configuration, and the plugboard guess is correct:

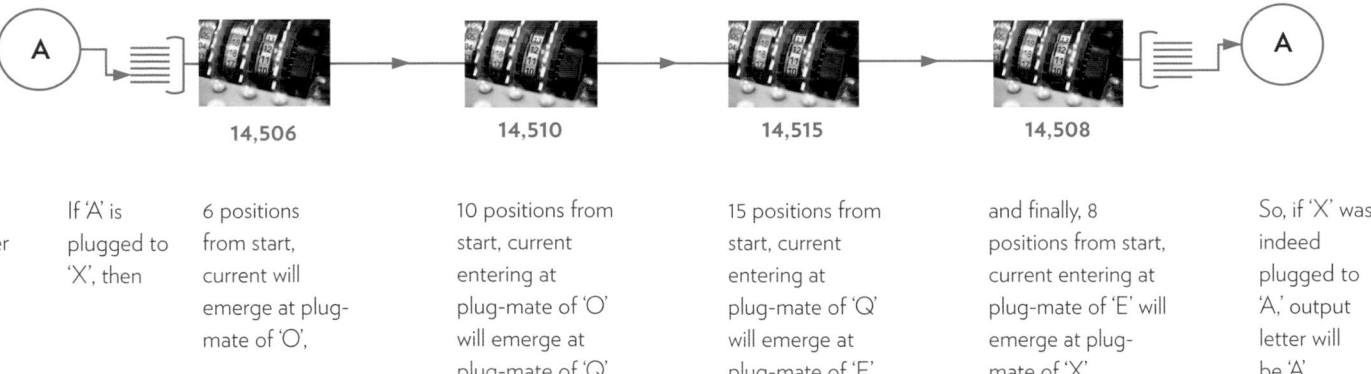

14,506   14,510   14,515   14,508

Test letter 'A'

If 'A' is plugged to 'X', then

6 positions from start, current will emerge at plug-mate of 'O',

10 positions from start, current entering at plug-mate of 'O' will emerge at plug-mate of 'Q',

15 positions from start, current entering at plug-mate of 'Q' will emerge at plug-mate of 'E',

and finally, 8 positions from start, current entering at plug-mate of 'E' will emerge at plug-mate of 'X'.

So, if 'X' was indeed plugged to 'A', output letter will be 'A'.

If the drums are not in the right configuration, or the plugboard guess is not correct:

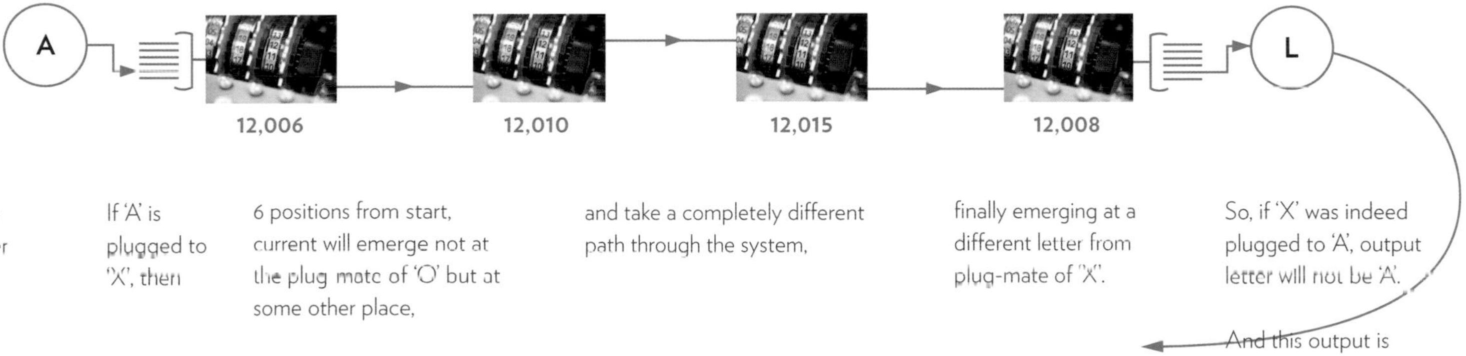

12,006   12,010   12,015   12,008

Test letter 'A'

If 'A' is plugged to 'X', then

6 positions from start, current will emerge not at the plug mate of 'O' but at some other place,

and take a completely different path through the system,

finally emerging at a different letter from plug-mate of 'X'.

So, if 'X' was indeed plugged to 'A', output letter will not be 'A'.

And this output is looped back to make a new input, which in turn will generate another false output, and so on.

A set of four Enigma scramblers are connected together in a loop. Electric current moves directly between them without needing to go through a plugboard. If the scramblers are in the correct relative configuration, at the end of the loop the same wire will be live as at the beginning of the loop. But if the configuration is wrong, the single perfect circuit will not be formed, and current re-enters the scramblers on another path. The Bombe senses the fact that multiple circuits are live and moves the drums round to test the next possible scrambler configuration.

Menus and spaghetti

When they set up the Bombe for each run, the Wrens were told to rotate each set of drums, so that the Bombe run would start with each drum-set in the correct position in relation to the message. Each Letchworth Enigma is set to correspond with an encipherment selected from the message, and the drums need to be moved round by the requisite number of places so that their starting position matches the position of the enciphered letter in the message.

The back of the Bombe is where the electrical connections took place according to the menu. Care and discipline were needed to prevent it becoming an unmanageable tangle.

Each Letchworth Enigma has an 'in' and 'out' connector, enabling the scramblers to be joined together to create loops.

The 'zero' position for drums on the Bombe was ZZZ. Each set of drums mimics the action of the scrambler of an Enigma machine at a particular point in the message. To get different sets of drums to test several encipherments from the same message simultaneously, the Wren operating the Bombe would rotate each set to the appropriate position. Taking the example on page 14, she would have to set up four Letchworth Enigmas for positions 6, 8, 10 and 15, offsetting the drums from their zero position by moving them to start at ZZF, ZZH, ZZJ and ZZO respectively.

The menu then told the Wrens how the Letchworth Enigmas should be connected together. The menu represents, as a wiring diagram, the hypothesis that the codebreakers wanted to test: that the crib does actually correspond to the intercepted cipher-text. Complex menus might have several loops and branches.

Round the back of the Bombe there is a spaghetti-like tangle of wiring. The cables are all 26-wire links which connect the drums together, but it's much harder to see what's going on than with the orderly rows of drums along the front.

Taking a close look at the back of the Bombe, you will see that there are several columns of electrical jacks, which have small black labels. These signify different parts of the Bombe and enable the connections to be set up.

- The third column from the left has numbered labels each stating 'in' and 'out'. Each number refers to one of the mock scramblers you saw at the front of the machine. The 'in' jack is where the input cable is connected, and the 'out' jack is where the output is connected.

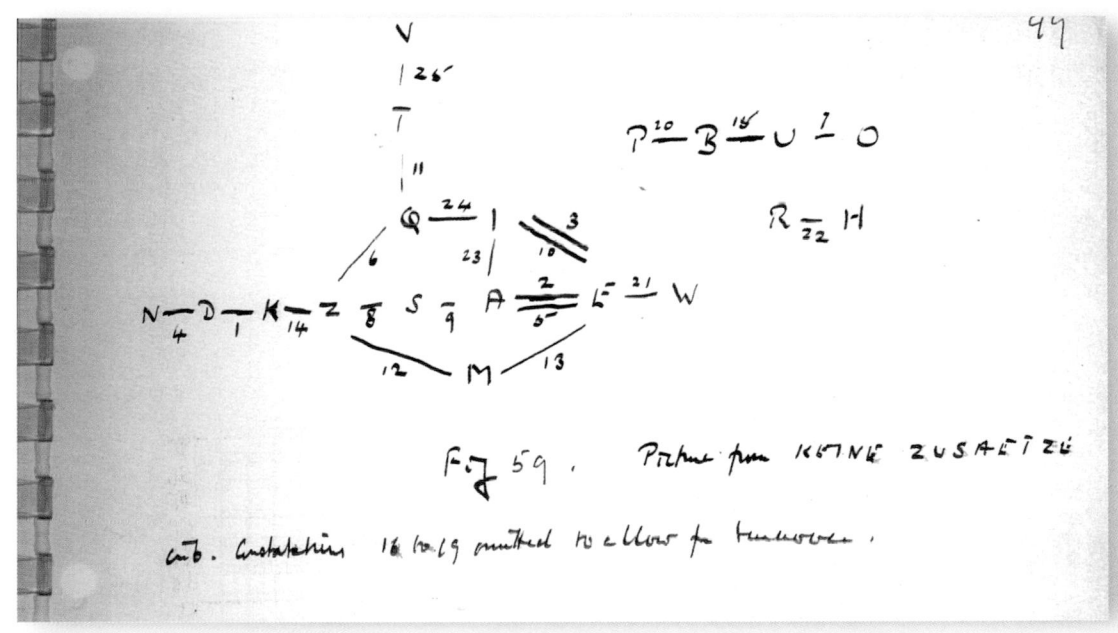

A menu drawn by Alan Turing and included in 'Prof's Book'. The legend reads: 'Fig 59. Picture from KEINE ZUSAETZE crib. Constatations 16 to 19 omitted to allow for turnover.' ('Constatation' is synonymous with 'encipherment'.) © Crown Copyright – Used with permission of Director GCHQ.

- Next, look at the second column from the left. This comprises a set of connectors called *commons*, labelled on the rebuilt Bombe 'CO1', 'CO2', etc. All the connectors labelled 'CO1' are electrically interconnected. So a cable plugged into CO1 is plugged into a junction-box, rather like an ordinary electrical extension lead with several sockets. If the cable connecting the output from scrambler 1 is connected to CO1, CO1 will enable the live wires in that cable to activate the equivalent wires in several more cables serving as inputs to other scramblers.

The commons are used where there are junction-points in a menu. Junction-points arise when there are several loops in a menu. Junction-points are where electrical connections are made between the Letchworth Enigmas, and they have a vital role in solving the problem of the plugboard.

Smelly and noisy

'You had to plait up this machine at the back with these great big leads which had to be plugged into different bits. Then at the front, you had this rack with rows and rows of drums marked up by colour and you were told what combination of colours you were to put on. You would set them all, press a button and the whole row went round once and then moved the next one on. It took about fifteen minutes for the whole run, stopping at different times, and you recorded the stop and phoned it through and, with any luck, sometimes it was the right one and the code was broken. It was very smelly with machine oil and really quite noisy.

The machine kept clanking around and unless you were very lucky your eight-hour watch would not necessarily produce a good stop that broke a code. Sometimes you might have a good day and two of the jobs you were working on would break a code and that was a great feeling, particularly if it was a naval code.'

Morag Maclennan, WRNS
quoted in Michael Smith's
Station X: The Codebreakers of Bletchley Park

Making a meal of it

The codebreakers created their menus from cribs, by looking for possible loops in the correspondence between the intercept and the possible plain-text. A menu can be constructed from the simple example on page 14 used to connect four sets of drums using 26-way cables.

We could also create a second loop from the same message using the positions coloured blue, as well as position 8.

Message position	1	2	3	4	5	6	7	8	9	10	11	12	13	14	15	16	17	18	19	20
Intercepted cipher-text	S	L	Y	A	R	O	P	X	J	Q	K	E	B	K	Q	Z	A	N	Y	G
Crib (possible plain-text)	K	E	I	N	E	X	B	E	S	O	N	D	E	R	E	N	X	E	R	E

The cottages where Alan Turing worked in late 1939 on his design for the Bombe.

This loop could be represented diagrammatically as follows:

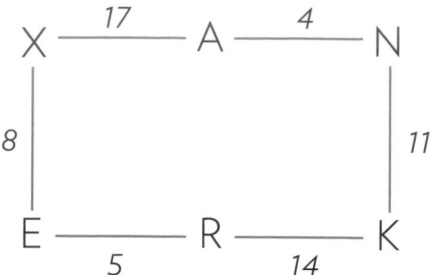

The loops could be joined together to form a figure-of-eight pattern, with letters 'X' and 'E' common to both loops. Electric current applied to letter 'X' at the input to the first set of drums should flow around both loops, provided of course that the crib is a good guess at the plain-text and the rotors are in the correct configuration.

A good menu has lots of loops, as this reduces the risk of a chance orientation of drums leading to a complete electrical circuit.

But a long crib contains a danger. At some point during any run of 26 letters, the middle rotor of the Enigma will have moved on one position, because the turnover notch in the ring of the right-hand rotor will have engaged. A turnover ruins the integrity of the crib: menu-making needs the rotors to be in the same relative configuration for the loop to close. Short sections of message are ideal.

The positions in the menu identified by letters 'X' and 'E' are junction-points between loops in the menu. They are also places where more than two Letchworth Enigmas can be connected with cross-and-connect plugs or 26-wire cables using the commons we examined on page 24. Having the option to apply a voltage to any of the 26 wires – not just the 'X' wire – at a junction-point is a feature of the Bombe which facilitates an attack on the Enigma's most formidable challenge: the plugboard.

Nothing to report

'One type of crib was obtained from the repetitious content of routine daily reports, often passing between low-echelon commanders and their immediate superiors. We developed a very friendly feeling for a German officer who sat in the Qattara Depression in North Africa for quite a long time reporting every day with the utmost regularity that he had nothing to report. In cases like this we would have liked to ask the British commanders to be sure to leave our helper alone.'

Gordon Welchman, *The Hut Six Story*

Other real Bletchley menus

'Personally I could not swallow the sponge puddings which emerged from long metal containers. They were pale yellow (bogus vanilla), bright pink ("rose" scent flavour), mauve ("violet" scented – ugh) and brown (cocoa – and just edible if one was really hungry). The "custard" sloshed over them helped to get them down to some extent.'

BP veteran quoted by Marion Hill

'First of all on the counter were rectangular dishes containing salads. Many of these, to the young British person who had never been abroad, were exotic to a degree: sweetcorn, olives, Russian and Waldorf salads, all kinds of mixed vegetables and dressing, surprising to the eye accustomed to a salad comprising a few lettuce leaves, a sliced tomato and some beetroot.'

Gwen Watkins, WAAF

'Our canteen outshone any sleazy restaurant. The smell of watery cabbage and stale fat afflicted us to the point of nausea. One day I found a cooked cockroach nestling in my meat.'

BP veteran quoted by Marion Hill

One in 150 million million – the plugboard

So far, we have looked at the problem of the rotors in the Enigma's scrambler. Statistically, of course, the plugboard is a challenge approximately 10^{10} times as difficult.

In a real Enigma machine, electric current reaches the first rotor by passing through the plugboard after the operator presses a key on the keyboard. The Bombe doesn't exactly have a plugboard, but it does allow its operator to choose a possible cross-plugging for a selected letter of an intercepted message.

Take another look at the Bombe, this time the narrow end of the machine. You will see these features:

- Four columns of 26 switches. The three towards the left are connected to the electrical input on a chain of drums. The chain is the connected-up collection of Letchworth Enigmas, with their drums moved to correspond to the relevant position in the message. Throwing one of these switches determines which wire will be made live in the input cable, which feeds current into the Letchworth Enigmas. The input cable connects to one junction-point in the menu. By this link, the panel of switches connects the scramblers to the control panel, in a way that is analogous to one of the plugboard connections between the rotors of an Enigma machine to its keyboard and lamps. A sense relay, which can detect the absence of live voltage on its wire, is also connected to each of the 26 wires in the input/output cable. If no current reaches a wire, the relevant relay 'sets' and the Bombe will stop.
- A rather unassuming prism-shaped device with a narrow horizontal window. This is the indicator panel that tells the Bombe operator, when the Bombe stops, which wire or wires are live in the input/output cable. There are two possibilities: it could be that a single wire has remained live, which is the case we have been considering so far, where the electricity runs around in a perfect circuit. There is

The control panel at the side of the Bombe. The columns of switches are for testing the plugboard hypothesis, and the horizontal device alongside shows the plugboard pairing output when the Bombe stops.

also another case, though: the read-out will also tell the operator if a single wire is not live when all the remaining 25 wires are live. The 25-live-wire case is just as important – and 25 times as likely – as the single-live-wire case, and is explained on page 35. Where the Bombe is testing several rotor orders or menus at the same time, the panel also tells the operator which chain of drums caused the Bombe to stop.

The control panel at the end of the Bombe is the interface between the operator and the Letchworth Enigmas cabled together according to the menu. But it is also the means by which a single wire at a selected junction-point will be made electrically live, and mimic the effect of the plugboard for a single letter.

Not just maths

It's tempting to focus on how the codebreakers used their mathematical training to develop the algorithms to crack open the Enigma. But the engineering behind the Bombe was state of the art as well. The fast drum on the Bombe cycles at around 100Hz, testing all 26 letters in about a second before the middle drum clacks onto its next test position. Each time all the electrical connections precisely line up on all drum-sets in the chain with no shorting. And the Bombe stops whenever electricity fails to reach all 26 wires in the input/output cable. In 1939, nobody had built a machine to do something like that before.

The Bombe became a reality through collaboration between the British Tabulating Machine Company in Letchworth and Bletchley Park. Without the persistence and imagination of BTM's engineer, Harold Keen, the Turing-Welchman vision for a Bombe would have remained just a paper concept. Keen and BTM delivered prototypes to Bletchley Park by summer 1940 and subsequently moved over to factory-style production.

Keen was known as 'Doc' to his colleagues, including those at Bletchley Park. His son John explains: 'As a young engineer he was spending much of his time travelling between home,

Doc Keen, the engineer who brought the paper concept for the Bombe to life.

the engineering works and customer installations and carrying paperwork, tools and minor technical items. To carry these he used a small case similar to, or possibly an actual, doctor's bag, then in common use. Because of this habit his fellow engineers came to refer to him as "Doctor Keen".'

Nicknames often stick because they hold a fundamental truth. Doc Keen's ability to anatomise and diagnose the ills of his early electromechanical patients, and then to carry out the precise surgery needed, was exactly what the mathematicians and intelligence officers needed to complement their own skills.

Doc Keen's contribution was fundamental to the successes achieved at Bletchley Park.

Nerve centre

The attack on the plugboard begins by selecting one junction in the electrical diagram constituted by the menu. The codebreakers called this junction-point the *central letter* of the menu. The central letter was usually the point in the menu where the largest number of loops and branches began – the busiest junction. Here the input voltage would be applied to a single wire in the 26-wire cabling connecting the Letchworth Enigmas.

In an Enigma machine, the connection between the scramblers and the outside world is made via the plugboard; in the Bombe it is made via the control panel at the end of the machine. Both machines have only one way in and one way out. If we were to suppose that the Enigma operator had cross-plugged 'X' to 'A', the electric current would still flow round a single circuit so long as we apply the voltage to the 'A' wire in the input cable instead of the 'X' wire. At the end of the circuit, the voltage would make only the 'A' wire live: the indicator panel would only show the letter 'A' as active, and that would confirm the hypothesis that 'X' and 'A' were cross-plugged.

So if 'X' has been cross-plugged to 'A', we should not apply a voltage to the 'X' wire at the junction-point where letter 'X' appears in our menu; we should apply it instead to the 'A' wire. We can do this on the control panel by flicking the 'A' switch, to make the 'A' wire live at the junction-point in the menu where current enters the banks of drums; or, to express it in other words, to test whether 'A' is cross-plugged to 'X', which is our chosen central letter.

And (just as with the test for the rotor configuration) if electricity spills out all over the place rather than staying within the confines of a single circuit, that would indicate that the hypothesis that 'X' and 'A' are cross-plugged is wrong.

Here the Bombe operator has decided to test whether 'N' is plugged to the central letter of the menu. The 'N' wire will be live in the input cable connected to the jack at the back of the Bombe, which maps to the central letter in the menu. Operators were instructed to select as the test-letter a letter from the menu at least two links away from the central letter.

There is one vital exception to that principle, which is the more likely 25-live-wire outcome. This is one of the clever features of the Bombe's design: if in the indicator panel a different letter is shown – say 'G' in our example – that means that no electricity reached the 'G' wire in the input/output cable. And that suggests that the drums are actually lined up correctly, but that 'X' was actually plugged to 'G' instead of 'A'.

Testing the plugboard hypothesis

1   The codebreakers first constructed a menu.

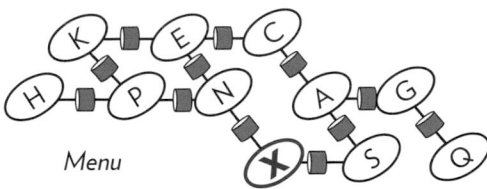

Menu

2   The menu relies on the plain-text and cipher-text, both of which will be transmuted by the plugboard as the current passes to and from the scrambler. In the Enigma machine, the current would enter the scrambler at the wire corresponding to the *plug-mate* of the keyed-in plain-text.

3   Each letter in the menu could have been transmuted into anything – and the Bombe caters for all 26 possibilities. Each menu letter thus represents a 26-wire connection point between mock-up scramblers. The wires at each point represent the plugboard options for the corresponding menu letter.

4   The mock-up scramblers on the face of the Bombe are connected using the menu as a map. A voltage can be applied to any wire at any connection point.

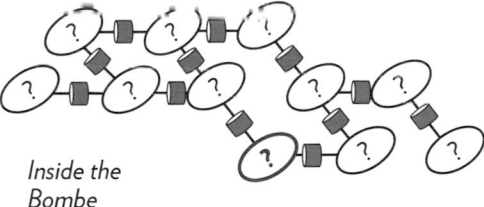

Inside the Bombe

5   Because none of the plugboard partners for the menu letters are known, all connections between Bombe drums are made using 26-wire cables.

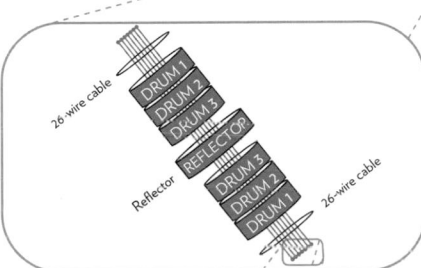

6   The Bombe tries out a hypothetical plugboard partner for just one letter in the menu – in this example, the Bombe will test whether 'X' in the menu is plugged to 'A'.

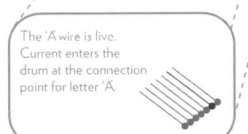

The 'A' wire is live. Current enters the drum at the connection point for letter 'A'.

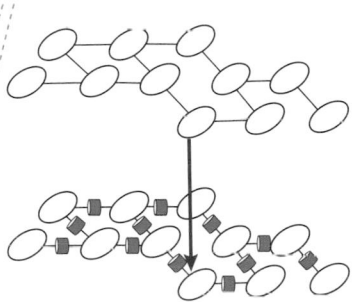

7   At the connection point in the menu which is labelled 'X', the 'A' wire is made live by applying a voltage.

Anatomy again

Time to have another look at the back of the Bombe. There are some columns of jacks which have different labels from those we examined before.

- Looking across to the right, one set of jacks has different labelling from all the others: the last column but one on the right. Here is where the cabling and scramblers are connected to the input-control and, on a stop, to the read-out panels at the end of the Bombe. The top three jacks are labelled 'CH1' to 'CH3', corresponding to chain 1 to chain 3, which you can also see printed on the control panel. A chain is an assembly of Letchworth Enigmas connected together: so the Bombe can (if there are enough sets of drums available) test three menus simultaneously; or, more commonly, test three different rotor-orders, each set up on a different chain with the same menu.

The plugboard hypothesis which the operator is going to test – that the central letter of the menu is perhaps cross-plugged to letter 'A' – will be fed in via a cable connecting the central letter to the chosen chain on the control panel. Typically, the cable from the 'CH1' jack will lead to a connector, which via other cables spreads the current out to the scramblers specified by the menu, and also to another hugely important component of the Bombe – the *diagonal board*.

- The first column of jacks on the left is neither diagonal nor a board, but it is actually a diagonal board. The jacks are labelled 'Z' through to 'A'. Each jack can be connected via a 26-wire cable to a connection point on a menu: in other words, the labels 'Z' to 'A' correspond to the letters in the menu. The wiring behind the diagonal board connects the 'Y' wire from the 'Z' jack to the 'Z' wire in the 'Y' jack and so on, as described on page 38.

Here we have the two fundamental design elements of the Turing-Welchman Bombe: testing for logical consistency using electrical feedback loops, and the diagonal board, which multiplies the number of possible loops for simple menus.

Ruth Bourne, who as a Wren operated a Bombe from 1943 until the end of the war.

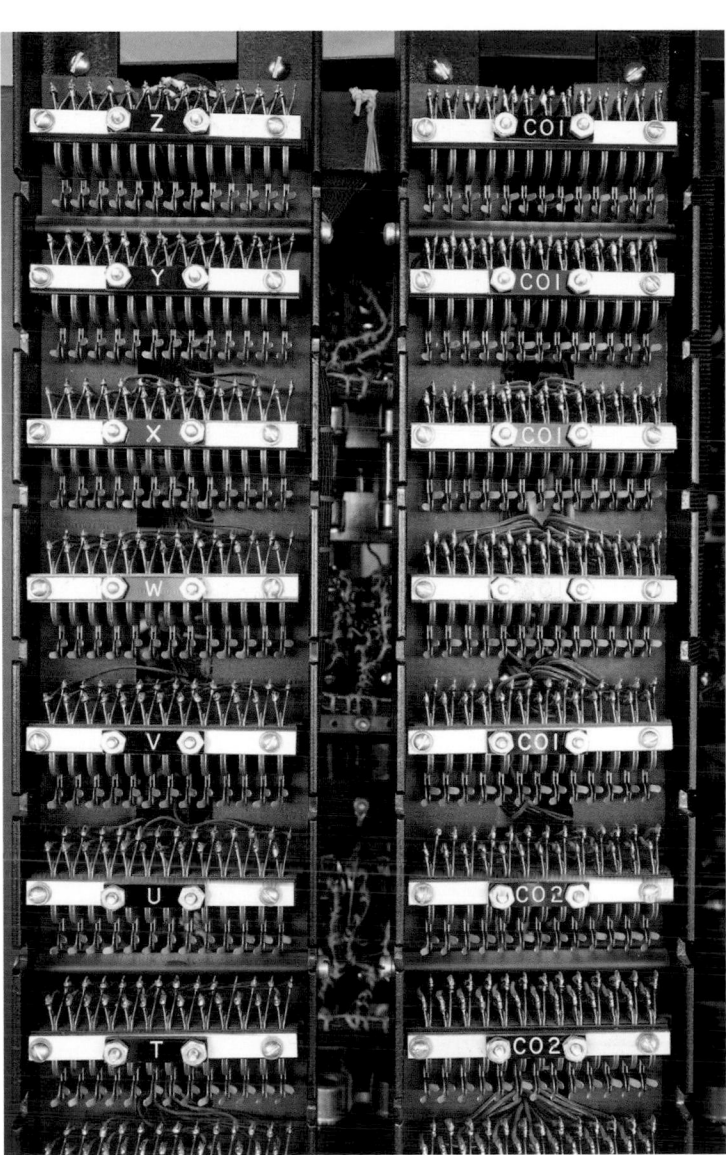

The front of the diagonal board (left column) and a set of commons (right column). The letters on the jacks on the diagonal board correspond to the letters in the menu.

(2)

OPERATION OF BOMBE SECTION.

The time taken to put on a new job varies considerable, but all men in charge of watches know from experience the quickest and most efficient course to adopt.

Plugging up varies from 5 to 35 minutes, depending on the menu and machine. The easiest machines to set up being "Agnus" and Wavendon. Due to the layout of the other machines careful consideration is required when preparing a menu for plugging up, bearing in mind the following points:-

1. Keep the number of commons to a minimum ; the less the number of connections the higher the conductive efficiency.

2. Prepare set up so that the operators can set all their banks the same, provided the job is on three times. This necessitates cross plugging where 10 OUT and 11 IN on chain 1 and 2, and 8 OUT and 9 IN on chain 3, are not adjacent. This must be allowed for in the set up to overcome contact trouble. Choice of input varies with the menu involved on each machine.

Time taken in drum changing varies on each machine. Agnus and Ming being the easiest. Jumbo takes longer due to having to reset on account of German Ringstellung. The screw on operation of Bonzo is very slow and laborious.

Operators try to arrange wheel orders to avoid changing of fast drums, preferring the slow drums, with centre drum as second choice. The reasons are:-

1. Once the fast drums are set they cannot get out of position. Therefore unless changed do not need checking. Being all of various settings they are most awkward to check and change due to their position on the machines.

2. The lower drums are usually set the same; so can be quickly checked, they are the easiest to change.

3. The centre drums are the easiest to get out of time, so must be checked each run.

An extract from a document of May 1942 explaining several operating difficulties with the four Bombes in Hut 11. © Crown Copyright – Used with permission of Director GCHQ.

Alan Turing in his parents' garden in the mid-1930s.

Turing's machine

The Bombe's principal design features were invented by two of Bletchley's most original thinkers: Alan Turing and Gordon Welchman.

The basic idea behind Alan Turing's Bombe design was to use cribs that involved loops of letters which could be used to create electrical loops within the cabling. To test the hypothesis that the 'central letter' of the menu is plugged to 'A', for example, you apply a voltage to a single letter in the loop and if the drums are correctly lined up, and the hypothesis on cross-plugging is right, then the current will flow neatly round in a circuit and only the wire for letter 'A' will be live in the input/output cable. But if the hypothesis that the central letter was plugged to 'A' is wrong, then the current enters the Bombe in the wrong place and it will emerge at some unpredictable place.

Turing's idea was to maximise the number of wires which become live whenever the plugboard hypothesis is wrong or the drums are in the wrong configuration. The more false outputs that come out of the test, the less likely it is that a stop could occur through coincidence.

There are in fact three possible outcomes. Continuing with our example of making the 'A' wire live in the 26-wire cable at the central letter junction-point of the menu:

1 The current comes back to the 'A' wire. That's great news because it means that the guess that the central letter was plugged to 'A' was probably correct. The Bombe will stop.
2 The current comes out everywhere, so that all 26 wires in the input/output cable are electrically live. This result, which is the most common one, means that the test has failed, and the Bombe drums move on round to try the next scrambler configuration.
3 The current comes out everywhere *except* one single wire in the input/output cable has zero voltage. This is the 25-live-wire situation. When electricity cannot reach a wire, the Bombe has found an

Is the Bombe a 'Turing machine'?

No, it isn't. A 'Turing machine' is a theoretical form of multi-purpose stored-program computer as described in Alan Turing's 1936 paper 'on computable numbers' (a copy of which is in the Bletchley Park Museum). The Bombe is a single-purpose machine which has to be re-wired every time to run its program.

After the war, Turing went on to work in various laboratories designing machines with full programmability. Among them was the one at Manchester University, headed by Turing's mentor and tutor Professor Max Newman. Newman had worked on the Colossus machine, which broke the Lorenz codes at Bletchley Park, and which some believe to have been the first electronic computer. It's no coincidence that these two towering figures were working on a real Turing machine.

alternative possible plugboard partner for the central letter: not 'A', but something else. Again, in this situation, the Bombe will stop. The 25-live-wire case can be illustrated with our example from pages 29–30. The wire with no voltage is the wire for one letter at the central letter junction-point of the menu. In our example that was the letter 'G'. When no electricity comes through the 'G' terminals of the mock scramblers connected up next to the central letter junction-point, but electricity is flowing everywhere else, there may be a complete electrical circuit within the Bombe which no voltage was able to reach. That implies that if, instead, we had made the 'G' wire live in the input cable, the indicator panel would have confirmed 'G' to be the letter plugged to 'X'. This last idea is very valuable: although we set up the Bombe to test only the one hypothesis that 'A' is the plug-mate of the central letter, the Bombe can find the actual plug-mate if the false outputs aim to cover the whole alphabet.

Going round in circles

Turing's feedback idea is that when the Bombe is testing a false plugboard hypothesis or a wrong rotor configuration, the current coming out of the loop in the wrong place can be used as another input, which itself is wrong too, and that will produce yet another wrong output, which can be fed back in and so forth. Whenever all 26 wires become live in the input/output cable, the Bombe would know it's time to move on to the next scrambler configuration. (Apart from that one exceptional case: if all wires *except one* are live, there's a good chance that the reason current could not reach that one wire is that it – or more accurately, the letter it represents – is the true plugboard partner of the central letter.)

Turing's feedback idea allowed the Bombe to test not only the simple proposal that a test-letter is plugged to the central letter (true or false), but to test simultaneously what *other* letters the central letter cannot logically be plugged to (all false), and thus to reveal the likely plugboard pairing of the central letter by a process of elimination. A false output from the circuitry informed the Bombe to continue with its run.

The trouble with Turing's original Bombe design was that it produced stops too frequently. Stops occur whenever the scrambler configuration and the menu produce a logically plausible result. The simpler the menu being used, the greater the likelihood that the stop was random and incorrect. Having several loops in the menus increased the amount of feedback and therefore cut down the number of false stops. But this approach depended on being able to create menus with several loops of letters, which was not possible often enough to make Turing's design functional.

Welchman, who had also been studying how machines could be used to solve the Enigma puzzle, invented an enhancement to the design which vastly increased the amount of feedback that could be achieved with relatively simple menus, and menus which did

Bombes at Letchworth joined together for a special decryption job.

Porgy, Alan Turing's teddy-bear.

not even have complete loops. Turing would later calculate that for a menu of 12 letters, the effect of the enhancement was approximately equivalent to adding two loops. And it would allow the codebreakers to split long cribs into short segments to reduce the risk of a middle-rotor turnover.

Welchman's idea was the diagonal board.

Another page from 'Prof's Book', where Turing explains how the diagonal board improved the performance of his Bombe design. © Crown Copyright – Used with permission of Director GCHQ.

Welchman's diagonal board

Welchman reasoned that the plugboard was reciprocal, so testing the proposition that 'A' is plugged to 'X' is the same as testing the proposition that 'X' is plugged to 'A'. In Turing's basic design, a connection was made only between 'A' and the central letter connection point.

At its most basic, Welchman's idea of reciprocity enables us to do this: as well as making the 'A' wire live at connection point 'X' in the menu, if the menu contains the letter 'A' as well, we could make the 'X' wire live at connection point 'A' too.

But Welchman's design was smarter than this. If the original input hypothesis ('A' is plugged to 'X') can be inverted, so can all the false outputs. Turing's design took the false output from the drums and made it another false input: in a case where 'A' is not the correct plugboard partner of the central letter 'X', the current might come out at, say, 'L'. If, at the end of the circuit, the 'L' wire in the cable at connection point 'X' has become live, this is simultaneously testing the hypothesis that 'L' is plugged to 'X'. Welchman saw that you could apply the reciprocity principle here too: why not also try the idea that 'X' is plugged to 'L'? What's more, this idea can apply to all the false outputs generated from each of the new false inputs, to create an entire cascade of false inputs across the whole alphabet.

The reciprocity was achieved through what, in engineering terms, was relatively straightforward cross-wiring in a device called the diagonal board. The diagonal board connects the 'L' wire in the 'X' jack to the 'X' wire in the 'L' jack; and likewise the 'A' wire in the 'B' jack is connected to the 'B' wire in the 'A' jack, and so on throughout all 325 possible pairings. This meant that if any wire became live, there were up to 25 more possible false inputs that could be used to rule out that particular configuration of rotors and plugboard hypothesis.

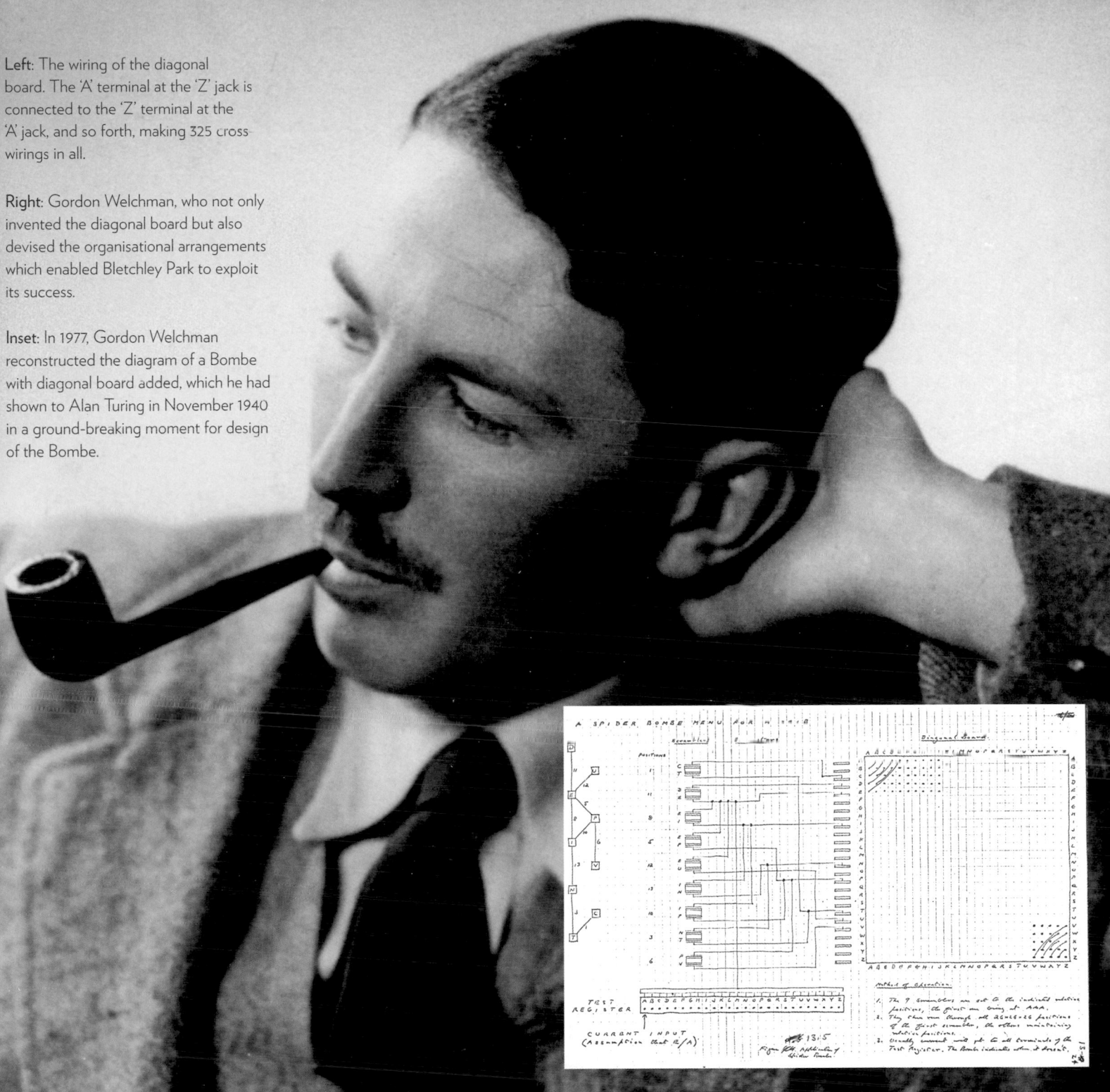

Left: The wiring of the diagonal board. The 'A' terminal at the 'Z' jack is connected to the 'Z' terminal at the 'A' jack, and so forth, making 325 cross-wirings in all.

Right: Gordon Welchman, who not only invented the diagonal board but also devised the organisational arrangements which enabled Bletchley Park to exploit its success.

Inset: In 1977, Gordon Welchman reconstructed the diagram of a Bombe with diagonal board added, which he had shown to Alan Turing in November 1940 in a ground-breaking moment for design of the Bombe.

Each 26-wire connection point – each letter of the menu – can be connected to a jack on the diagonal board, so the diagonal board has one jack for each letter of the alphabet. The connections to the diagonal board maximise the spread of current across the system whenever the rotor configuration or plugboard hypothesis is wrong.

The beauty of the diagonal board was that it scaled up Turing's feedback design massively without adding appreciably to the cost of building a Bombe. Welchman's addition also allowed the testing of menus with dead-ends as well as loops, making menu design easier.

These design features were not imagined by the officers who had been tasked by the German Armed Forces Supreme Command to test the Enigma for flaws, and in consequence the Germans always believed that Enigma was unbreakable.

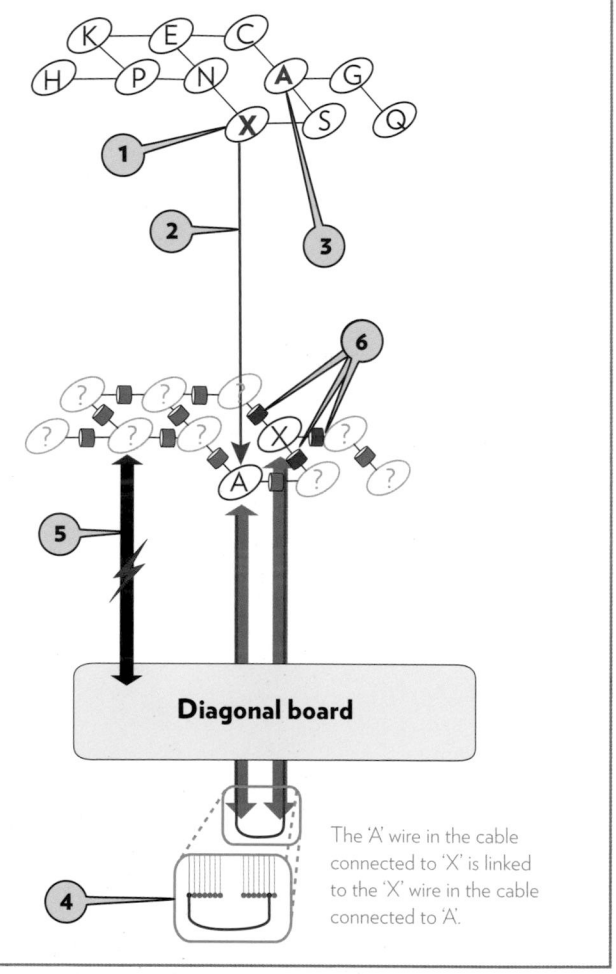

The 'A' wire in the cable connected to 'X' is linked to the 'X' wire in the cable connected to 'A'.

The diagonal board

1 Each letter of the menu corresponds to a connection point joining mock-up scramblers. The connection points have 26 wires, one for each input/output letter on a scrambler.

2 A live wire at a connection point is testing the hypothesis that the menu letter is connected to that scrambler input letter. So, in this example, the live 'A' wire at the 'X' connection point is testing whether 'X' is plugged to 'A'.

3 But testing whether 'X' is plugged to 'A' is the same as testing whether 'A' is plugged to 'X'. In other words, the 'X' wire could be made live in the 'A' connection point in the same menu.

4 The diagonal board provides reciprocal wiring for all possible letter pairs.

5 Each connection point in the menu can be plugged into the diagonal board to spread the current around the menu circuit.

6 In this example, the diagonal board connects the live 'A' wire at the position 'X' with the 'X' wire at position 'A', allowing current to enter three more mock-up scramblers at the connection point for 'A'.

Getting the read-out

The Bombe is designed to stop its run when it has found a plausible relative orientation of the three rotors and a plausible plugboard partner for the central letter of the menu.

The three special indicator drums on the right of the middle bank show the rotor configuration relative to the notional ring-setting ZZZ. The suggested plug-mate for the central letter appears on the small indicator read-out on the control panel at the end of the Bombe.

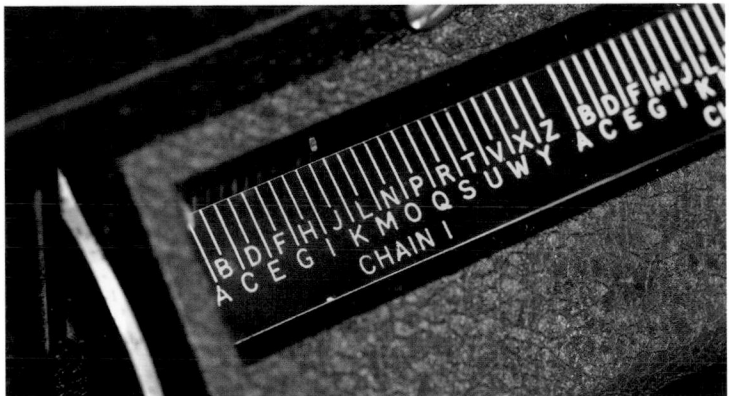

The prism-shaped indicator unit showed, with a small metal bar in the 'up' position (here at letter 'K'), that a single wire in the input/output cable was not live, thus revealing the possible plugboard pairing of the menu's central letter. (If instead the test-letter selected in the bank of switches in the control panel was a good guess, then voltage would fail to reach all wires except that one in the input/output cable; and 25 of the 26 indicators would be 'up'.)

The Wrens who had the noisy and tedious job of operating the Bombes would note down the stop position on the three indicator drums, and the plugboard letter from the read-out on the indicator panel. They would then restart the Bombe to continue its run, since the stop might just be a coincidence rather than the definitive answer

At a stop, the drums show the suggested position of the rotors on the Enigma for the starting point of the message being tested. This information is presented on the indicator drums as a set of offsets.

The three gold-coloured drums at the end of the middle row show the positions of the drums, relative to the starting position ZZZ, where the Bombe reached its stop.

- When the Letchworth Enigmas were set up on the Bombe, each set of drums was rotated from the 'zero' or ZZZ position.
- The drums don't have ring-settings like the rotors of an Enigma. Without ring-settings, the three drums will behave like rotors with rings also set to ZZZ. When the Bombe stops, the drums will have rotated around to something other than ZZZ, say NKV – a possible starting configuration, still assuming that the rotors' rings were set at ZZZ.
- The indicator drums present the information the other way round: as a set of offsets which would give the same encryption from

an Enigma machine showing the letters ZZZ through its viewing windows before the first keystroke was made. (For a stop of the drums at NKV, the rings of the Enigma rotors would need to be at MPE to get the same encryption with a starting configuration of ZZZ.) This is why the letters on the indicator drums are arrayed clockwise while the letters around the regular drums are arrayed anti-clockwise.

For the codebreakers, finding the actual ring-settings would have to wait for later. Meanwhile, the rest of the plugboard had to be solved.

Above: Wrens at work at Eastcote, one of the Bletchley Park outstations.

Below: The official GC&CS (Government Code & Cypher School) badge awarded to veterans of Bletchley Park war service.

The checking machine

The Turing-Welchman Bombe only manages to crack the Enigma settings in part. Assuming that the Bombe did not stop randomly – after all, it is only finding a plausible setting, and doesn't guarantee the correctness of its result – there are still a few outstanding problems to solve. The first one is to find the remaining plugboard settings.

To continue with our example, the Bombe operator has found the rotor configuration and a plausible plugboard partner of letter 'X'. But for most of the war the Germans used 10 cross-pluggings, so there are still 9 more to identify – about 4.6 million million possibilities. Those 9 plugboard settings might seem like a dauntingly difficult wall to break

The checking machine behaves like an Enigma machine without a plugboard, but enables the operator to move the drums around by hand.

Wrens in Block D using checking machines.

through, but the codebreakers used the simplest solution of all: they would get the Enigma machine itself to tell them the missing settings.

Enigma machines were in short supply at Bletchley Park, so BTM built another device, this time without a plugboard, called the *checking machine*.

Take a look at the rebuilt checking machine. It has the expected trio of drums, which will be the three drums used in the successful Bombe run. (On the left, there is a fourth drum, which could be used for tests relating to 4-rotor Enigmas.) Another difference between the drums used on the Bombe and those used on the checking machine is that the checking machine drums have adjustable bezels, like the rotor-rings in the Enigma scrambler. And the drums can be turned round quickly by hand, in either direction.

There's also a keyboard and a lampboard, just like an Enigma, but no plugboard.

The checking machine is wired up just like the scrambler unit in the Enigma, but the absent plugboard is, intriguingly, the feature which is just what the codebreakers needed to reveal the missing cross-pluggings.

The missing links

The checking machine works in the secret world behind the plugboard. Recall that when the Bombe was connected up, the connection points each corresponded to a letter in the menu; but those connection points were in fact the *plugboard partners* of the letters in the menu, since all plain-text has to go through the plugboard before it gets to the rotors in the Enigma machine's scrambler.

The Bombe has revealed the letter to which the central letter of the menu is cross-plugged. That one known cross-plugging provides a breach in the plugboard's security wall, which enables the remaining plugboard connections to be captured easily.

Let's follow the process through using the simplified menu on page 15. And let's say that when it stopped, the Bombe's indicator panel told us that the plugboard partner of the central menu letter 'X' is actually 'A', as in the diagram on page 21. So, in the behind-the-wall part of the Enigma machine, we know that the live input wire going into the first rotor was the 'A' wire when the Enigma operator pressed the 'X' key on his machine at position *Start* plus 6.

The checking machine – the Enigma behind the plugboard wall – is set up with its drums matching the stop given by the Bombe. First the bezels – the equivalent of the rings on a real rotor – are set to the

Block D as it is today. After it opened in 1943, this building housed the Hut 3 and Hut 6 teams, which created the menus and decrypted the German army and air force messages.

'The checking machines at the side of the Bombe were placed there only a short time before VE day, much to the disapproval of the checking Wrens. They preferred the quiet of the checking room.' Silvia Pulley WRNS, quoted in Gwendoline Page, ed., *We Kept the Secret*.

offsets shown by the Bombe's indicator drums. Then the checking machine drums are moved round to *Start* plus 6 (ZZF). We can then press the 'A' key and find out what letter the scrambler comes up with after the current has completed its double passage through all three drums and reflector. That letter – it could be anything, but let's say it is 'R' – will light up on the checking machine lampboard. And that output letter 'R' is what, on the secret side of the plugboard wall, should correspond with the plain-text letter in the menu at *Start* plus 6. From the menu we believe that 'X' became enciphered as 'O', so we have now got a plausible plugboard partner for 'O' as well.

Now we can move the checking machine drums on round to *Start* plus 10 (ZZJ), and press down the 'R' key, and the output should be the plugboard partner for 'Q'; and so on until all the connections in the menu have been tried out. If the menu was not long enough to obtain all the plugboard connections, the rest can be found out from the crib. Take a letter for which the cross-plugging is not yet known, but which has been enciphered somewhere in the message into a letter for which the pairing is known, and the checking machine can be set up at that position to reveal the unknown plugboard pairing.

Using the checking machine

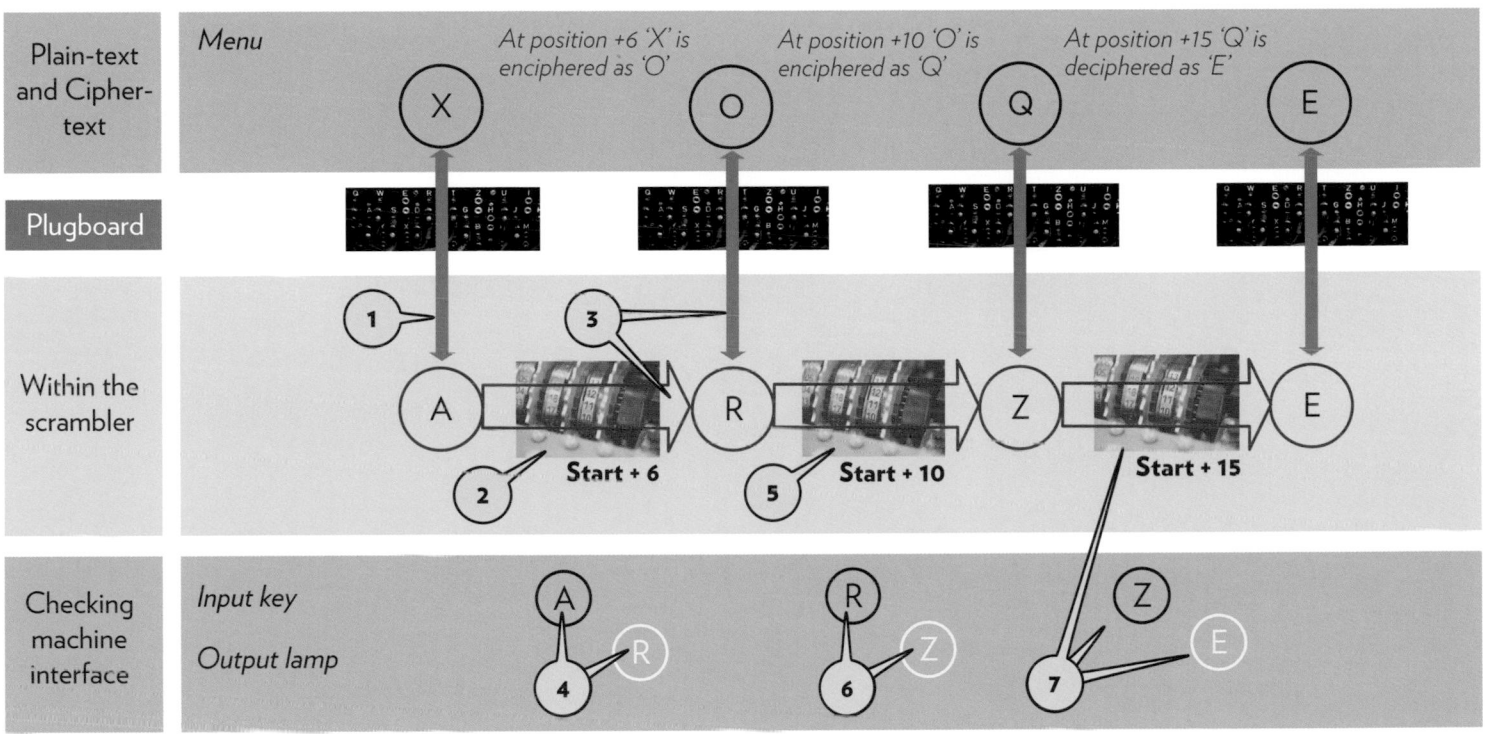

	Menu	At position +6 'X' is enciphered as 'O'	At position +10 'O' is enciphered as 'Q'	At position +15 'Q' is deciphered as 'E'

Plain-text and Cipher-text: X — O — Q — E

Plugboard

Within the scrambler: A — Start + 6 — R — Start + 10 — Z — Start + 15 — E

Checking machine interface — *Input key* / *Output lamp*

1. The Bombe has suggested that 'X' is plugged to 'A', as well as giving the suggested start configuration of the rotors.
2. The drums on the checking machine are moved round to 6 positions after the start position.
3. At this position, the letter 'A' will be converted by the scrambler to the plugboard partner of 'O'.
4. Letter 'A' is pressed on the checking machine, lighting up 'R' on its lampboard – 'R' is the partner of 'O'.
5. The drums are wound on to position 10, where 'R' will be converted to the plugboard partner of 'Q'.
6. Keying in 'R' at position 10 lights up 'Z'.
7. Then the drums are wound on to position 15, where keying in 'Z' lights up 'E', showing that 'E' is one of the 6 letters left without a cross-plugging.

Check-mate

Two special checking machine results should be mentioned:

- If a menu letter is one of the six letters left without a cross-plugging, the lampboard read-out would be the same as the menu letter.
- If the Bombe has unluckily come up with a false stop – that is, a chance coincidental configuration of the rotors and possible cross-plugging for the central letter which is not the actual Enigma setting for the day – the checking machine would produce contradictory results, for example by giving the result that both 'W' and 'P' are allegedly cross-plugged to the letter 'M'. That can't be true, so the check rules out the stop as anomalous, and the Bombe needs to be restarted on its run to find a better solution.

So, the checking machine has another useful function: that of eliminating false stops. But it has another function as well, which is to solve the one remaining challenge of the Enigma: the ring-settings.

The ring-settings determine the points in the message at which the middle and left-hand rotors move round one place. Because we now know the relative orientation of the rotors and the plugboard settings, we can also use the checking machine to decipher the rest of the message using those settings. After no less than 26 letters – assuming the message is long enough – the clear German which had been coming out will revert to twaddle. So we'll know that by that point a turnover must have happened, and we can assume that the middle rotor on the German operator's Enigma machine had cranked round one place.

Possibly that doesn't work: it could be a false stop, some nonsense ('quatsch') inserted as a decoy into the plain-text to confuse codebreakers, or a typo or interception glitch; or it could be that both middle and left rotors turned over simultaneously. But if moving the middle rotor on by one restores the message to German plain-text, we have found the ring-setting on the right-hand rotor. It only remains to find the setting on the middle rotor – which determines the turnover of the left-hand rotor – and that will happen eventually if we test every 26th letter in the message from the first turnover point onwards.

But with luck the ring-setting for the right-hand rotor has been found. And at Bletchley Park that was enough to enable the codebreakers to discover the remaining settings.

The Mansion looking idyllic in summer.

A menu from 1943. © Crown Copyright – Used with permission of Director GCHQ.

Why bother with ring-settings?

You might ask 'why bother', if we can decipher the rest of the message without knowing the ring-settings.

But one deciphered message is not going to provide a lot of intelligence, particularly if it contains only a boring crib such as 'nothing special to report'. What the codebreakers needed to know was how all the Enigma operators had set up their machines for the day, so that all messages in that key could be broken.

Each message had a unique starting position for the three rotors, which the sender would notify to the receiver, albeit in enciphered form: the *indicator* – see the inset box on page 51 for more about this.

What the Bombe tells you is the rotors' relative starting positions for the message which has been broken. To decipher the other messages received, it is no use setting up an Enigma machine with these starting positions. You have to know the actual starting positions for each message, and to get that you have to be able to read the information sent by the sender as to where the rotors were positioned for the start of the message – which was sent using the day's ring-settings.

So, to decipher the other messages, the codebreakers had to find the ring-settings.

Clonking

The Bombe has revealed the *relative* positions of the core wiring in the three rotors on the (unlikely) assumption that the rings are set at ZZZ.

The checking machine has revealed where the turnover notch on the right-hand rotor is actually located in the message being examined – which is the same as knowing how the ring has been set on that rotor. Therefore, the codebreakers knew the *actual* start position of the right-hand rotor as well as its ring-setting.

The codebreakers were now left with just the ring-settings on the left and middle rotors to find. They used the *indicator* (see inset box) of the message they were analysing to find these remaining ring-settings.

Knowing the actual right-hand rotor start position is the key to this. In the example set out in the inset box, the actual right-hand rotor start position of the message is 'E'. But the German Enigma operator also sent that information in enciphered form when he transmitted the

Geheim! Nicht ins Flugzeug mitnehmen!	OKH-Maschinenschlüssel A Nr. 39	№ 00014

	Datum	Walzenlage			Ringstellung			Steckerverbindungen										Kenngruppen			
o	31.	V	II	IV	17	09	02	KT	AJ	IV	UR	NY	HZ	GD	XF	PB	CQ	sfy	azy	zkq	bqi
o	30.	I	III	V	22	12	10	UE	PL	AY	TB	ZH	WM	OJ	DC	KN	SI	iuy	swz	omo	myj
o	29.	V	IV	II	04	01	25	WJ	VD	PO	MQ	FX	ZR	NE	LG	UC	BK	rui	kao	fqi	rwu
o	28.	II	III	IV	05	03	12	HR	TJ	LD	IO	CN	GX	QK	PZ	WS	AF	ioy	kjv	yko	fpz
o	27.	I	II	III	10	20	15	AQ	ZK	MU	GH	ST	LN	XY	IJ	BF	RV	ggf	jus	lrs	glc
o	26.	II	V	I	15	09	06	DS	UL	ZJ	OI	HN	FT	RK	YC	XQ	GB	orl	rht	ksz	ego
o	25.	V	IV	III	26	07	18	WA	QD	XS	UY	LG	JI	FB	HK	MT	CE	pfr	ijw	zgg	ygj
o	24.	III	I	IV	04	19	24	OH	XM	DJ	IL	VU	KG	QZ	BT	FR	AS	nbt	pvd	eqo	wyn
o	23.	I	IV	V	11	17	01	QJ	GY	SH	OX	ZB	PL	FA	WI	VK	ND	hhv	hhq	kul	hmf
o	22.	IV	I	II	21	11	17	CV	LE	KN	UH	YJ	TI	RB	FZ	PA	MO	jlw	vrh	vya	pbf
o	21.	I	V	II	06	21	10	JN	UX	YT	BG	DR	QC	KE	SP	HZ	LA	zit	jlc	jbl	pvi
o	20.	V	II	III	07	18	04	ZG	NW	SM	VY	XT	UR	OC	LB	AQ	HF	ctx	gns	xeg	nvo
o	19.	IV	V	I	08	09	22	IT	YK	BL	RZ	VP	FN	JW	QO	MS	AE	lyx	jua	zju	nss
o	18.	I	IV	III	26	16	11	BU	TS	VH	JL	WX	AY	KG	ZM	PD	NF	ize	ysj	skw	znr
o	17.	III	V	I	11	22	16	GY	JN	SF	KI	LB	QD	UX	CW	HR	MA	xvd	kkb	poi	fug
o	16.	V	I	IV	04	09	24	QL	EY	BG	MN	ZO	AW	TC	VX	FS	HP	afp	uah	tpn	npf
o	15.	II	V	III	03	20	14	JD	BM	XR	LG	PC	OF	ZI	YH	VK	EW	nfk	pvm	vue	cpr
o	14.	IV	I	II	25	12	15	BT	OW	SN	DA	ZL	VP	QX	UE	HR	MC	zgo	omz	pdf	xuq
o	13.	I	V	IV	07	18	05	IW	NB	XO	YS	AJ	MQ	VH	FT	UL	RE	zor	ccm	odl	ijs
o	12.	IV	III	III	19	03	21	CN	LG	IZ	DO	SE	VR	TQ	KM	JF	AX	eqk	whq	avo	zpf
o	11.	V	II	I	08	20	14	HV	FP	CM	AJ	OU	YB	WS	NT	GK	EZ	hvm	icd	nxo	yxk
o	10.	IV	V	III	21	08	03	IJ	XR	ZV	NT	CK	OU	EB	PL	MY	HD	bgd	xka	gsg	sgs
o	9.	III	I	II	14	16	06	LN	IK	HS	DB	TX	CG	WY	EV	OF	RA	myh	ncz	xvx	ees
o	8.	IV	III	I	09	18	14	RG	XU	WZ	AP	LF	IY	SQ	DO	VJ	HT	ooq	xeo	ocn	kde
o	7.	II	I	V	18	13	24	EK	RO	JX	WV	HS	QP	BZ	MU	TN	CA	fmc	mkh	lhe	tmq
o	6.	III	II	IV	23	01	17	DC	VG	OL	UA	EK	ZH	YX	PW	IM	RF	tlo	wbj	sre	kjd
o	5.	V	III	I	19	23	15	QP	DG	ZJ	NK	SB	IC	FT	ER	UV	HA	hnp	wla	shv	spd
o	4.	IV	III	II	26	04	03	MX	QO	HI	TB	GA	KP	LZ	CS	WJ	NV	clc	jdh	yoq	hwt
o	3.	V	III	II	01	02	23	EI	DY	PO	SJ	FN	LB	RK	GX	AH	CU	jty	bzy	kdh	asq
o	2.	I	V	III	16	07	02	ZO	IA	VM	CT	PX	YB	HU	SD	RN	EL	uqn	nsx	jqk	pzb
o	1.	IV	I	V	20	05	10	SX	KU	QP	VN	JG	TC	LA	WM	OB	ZF	sro	eej	fnz	szk

Above: A chart showing the rotors to use, their order, the ring-settings and plugboard pairings for each day. The *Kenngruppe* is information which tells a recipient which key is being used.

Right: Adjusting the ring-setting on a rotor changes how the inner core wiring is orientated relative to the numbers or letters around the rim, and determines the point at which the adjacent rotor turns over.

indicator: in his message it was enciphered as 'R'. And now, because the codebreakers knew what the third letter of the indicator was, in both plain-text and cipher, the indicator has provided a miniature crib.

If the checking machine drums happen to be set up like the Enigma operator's machine for the indicator message, with the correct ring-settings, the third letter of the indicator would encipher correctly – in the example in the inset box, 'E' would encipher as 'R'. As an alternative to hand-methods, codebreakers were able to use the checking machine to test all of the 676 possibilities for the two remaining ring-settings. The question was how to get through all 676 without it taking an unacceptable amount of time. For this, the codebreakers used a technique called *clonking*.

The test involves trying to find ring-settings for the left and middle rotors where 'E' enciphers as 'R'. The letter 'E' (or rather, its plugboard partner) can be held down continuously on the keyboard of the checking machine, while a drum is rotated quickly by hand through its 26 possible positions. If you ask for a demonstration of this, you can see the lamps lighting up in succession as the drum is rotated. At various positions when the checking machine operator is clonking through the 676 permutations, the lampboard will light up 'R' (or rather, its plugboard partner), and each such position is a good candidate for the ring-settings in use for the day.

There would also be some false positives, but these could be weeded out by trying to decipher the rest of the indicator message and seeing if the result is consistent with the rotor configuration provided by the Bombe.

Once the last two ring-settings are known, all messages can be deciphered.

Message indicator procedure on a 3-rotor Enigma

The sender would choose at random – actually, operators being human, it sometimes wasn't random – three letters which would be the starting point of his enciphered message. These three letters, called the 'message setting', would show up in the little windows on the inner metal lid of the Enigma machine when the rotors had been twiddled round to the correct point. Let's say he chose UTE (because Ute was the name of his sister). Army rotors have numbers round the rim, so this meant the numbers 21, 20 and 05 would show through the windows.

Now the sender would have to tell the receiver that this is his start point. Rather than sending these letters in plain-text, the sender enciphered them. But here's a chicken-and-egg situation: to encipher UTE, how should the sender set up his machine? The Germans used several procedures for this, but after May 1940 the sender chose three more letters at random, called the 'indicator setting'. Let's say that our sender chose KLA (the first three letters of his girlfriend Klara's name); now he was ready to go. He set the rotors to show 11, 12 and 01 – corresponding to KLA – and typed in UTE, which came out as some sort of gobbledegook like ZQR. He reset his rotors to 21, 20, 05 (for UTE) and started enciphering the actual message. Along with other information in the preamble of the message, the sender would transmit the letters KLAZQR, a sequence known as the *indicator*. The indicator enabled the receiver to set his rotors to 11, 12, 01 (KLA) and type in ZQR; this lit up the letters UTE, so the receiver now set his rotors to 21, 20, 05 (UTE) and was ready to decipher the actual message.

Typex at Station X

Once the whole Enigma set-up had been found – the rotors and their order, the ring-settings, and the plugboard – this would enable any person with an Enigma machine to decipher an intercept by following the same procedure as the German Enigma operator receiving the same message.

But Bletchley Park wasn't on the German distribution list for supply of Enigma machines. Despite that, machine-based cryptography was something the Allies had also been developing, and the British had a machine called the Typex which could be modified to behave like an Enigma.

So at Bletchley Park there were typing pools equipped with Typex machines, where the operators could follow the German indicator procedure to find the starting position for each message. Once the

Typex machine was set up, the intercepts were keyed in and converted back to German. The Typex didn't have a lampboard – instead, its output was printed on a long, thin strip of paper, which could then be pasted onto a more useable sheet of paper and sent on for assessment of its value as intelligence.

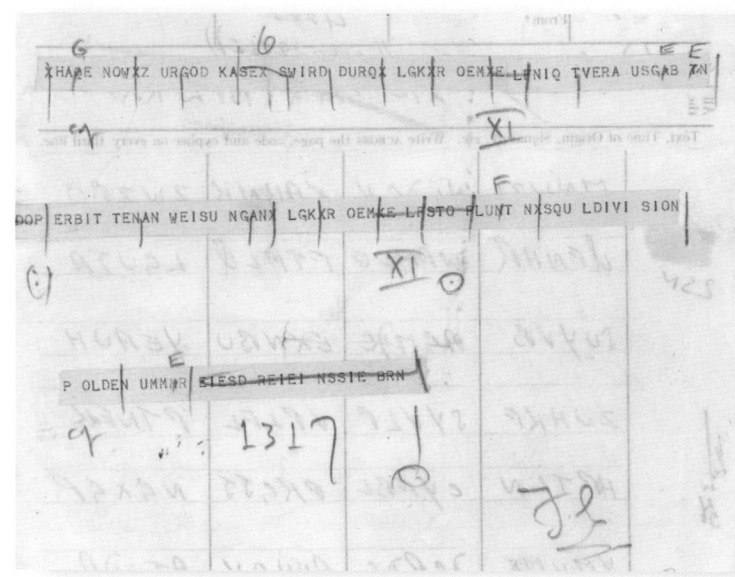

A decrypt generated by a Typex marked up to show the actual German. Note how the German cipher-clerk used 'X' as a space and 'Q' to represent 'CH', and at the end of the message how numbers had to be spelt out in full. The message is also full of typing or transmission errors. © Crown Copyright – Used with permission of Director GCHQ.

A Typex machine.

Bletchley Park staff using Typex machines in Block E.

What's a Typex?

'In 1926 an Inter-Departmental Committee was set up to reconsider the possibility of the use of cypher machines ... After a prolonged study of the various machines available, including the German Enigma ..., the Committee recommended in January 1935 that the Air Ministry should arrange for the construction of "3 sets of cypher machines of an improved 'Enigma' type through the agency of so-called 'Type X' attachments." This historic decision was the genesis of the secure Typex machine which was to be used successfully by the three Services and other government departments for many years.'

Sir Harry Hinsley, Bletchley Park 1939–45, *British Intelligence in the Second World War*

A decrypted message, sent two days before the Allied landings in North Africa in Operation Torch. The sender passes on intelligence that 50 bombers and 12 troop-carriers are heading for the Gibraltar straits, but warns that his informant may be unreliable. © Crown Copyright – Used with permission of Director GCHQ.

Coups and cuckoos

The intelligence which came out of the Enigma decrypts was used for many purposes. Sometimes it led to spectacular successes, and sometimes there were sadly missed opportunities.

Some that were coups

The *Bismarck*. On 24 May 1941 HMS *Hood* was sunk in the North Atlantic. Only three men of her complement of 1,418 survived. Her aggressor was the *Bismarck*. Outgunned and damaged, HMS *Prince of Wales* withdrew from the fight. HMS *Suffolk* had been shadowing but lost contact in the night. Honour demanded retribution, but where had the *Bismarck* gone?

The *Bismarck* was herself damaged and needed to head for port. The options were Norway or France, but the navy would have to plump for one, and if they made the wrong choice she would escape scot free. Various pieces of intelligence – though not much from Bletchley Park – suggested that France was the destination. There were no naval Enigma decrypts on the subject, although signals traffic analysis from Bletchley indicated that control of the big ship had shifted from Wilhelmshaven to Paris.

Until 1815 hours on 25 May. Just after Harry Hinsley put the phone down after another row with the Admiralty about the *Bismarck*'s

Bismarck firing at HMS *Hood* during the battle of the Denmark Strait.

destination, an Enigma decrypt of a message sent to Athens solved the puzzle. General Hans Jeschonnek, the German air force Chief of Staff, was planning his invasion of Crete. But he had a relative aboard the *Bismarck* and had asked about him. The response was that the heavy ship was heading for the safety of Brest. France, not Norway, it was.

The *Bismarck* was sunk by HMS *Rodney*, *King George V*, *Norfolk*, and *Dorsetshire* after a single torpedo fired from a Swordfish aircraft flown from HMS *Ark Royal* had damaged the *Bismarck's* steering gear. One hundred and ten survivors – destined to sit out the war as prisoners – were picked out of the water before the fear of a U-boat attack forced the rescue effort to be abandoned. The end of the *Bismarck* also meant the end of surface-raiding as a core part of German naval strategy.

Alam Halfa. General Bernard Montgomery arrived in the Western Desert as commander of the Eighth Army on 13 August 1942. His formidable enemy was Erwin Rommel, whose Panzers seemed to be able to sweep all before them.

Ultra intelligence derived from German army Enigma was reaching the British Forces in Egypt within 24 hours of interception. On 15 August, Rommel sent a fateful report to Berlin. He said that the Alam Halfa ridge was the key to the whole El Alamein position, and that he planned to attack before the British began to construct fixed defences. Bletchley Park passed this on to Monty's HQ; a couple of days later they reported that the attack would probably happen on 26 August; and over the coming days they produced a stream of data on Rommel's fuel and supplies situation, his physical health, and his appreciation of the Eighth Army's apparent state of unreadiness.

But behind the camouflage nets Monty was ready. He deployed his armour and artillery in a defensive line around the ridge, holding them back as if they were a stationary battery. Rommel began his attack

Eighth Army General Sir Bernard Montgomery during a visit to England in 1943.

Arctic Convoy PQ18 at sea in September 1942. An underwater detonation erupts between HMS *Eskimo* and the convoy's merchant ships.

on 30 August, but he could not breach the wall. Geography and fire-power were too much, and he ordered a withdrawal.

Monty had got his first victory over Rommel, and thus began the handover of initiative from the Axis to the Allies in North Africa.

And some that were cuckoos

PQ17. Convoy PQ17 sailed on 27 June 1942 bound for Archangel. They had to run the gauntlet of the *Tirpitz*, which was in Trondheim and

threatening to go to sea. Also lying in wait were U-251, U-255, U-376, U-408, U-456 and U-703. Among their commanders were Reinhart Reche, Max-Martin Teichert, Heinz Bielfeld and Heinrich Timm, whose careers between them notched up eight Iron Crosses, three Knight's Crosses, two German Crosses in Gold, and over 150,000 GRT of merchant shipping sunk.

Based on German air force and navy Enigma messages, there was good intelligence on the dispositions of U-boats and the German capital ships in the Arctic theatre. Bletchley Park reported to the Admiralty that the U-boats were homing in on PQ17, and that the *Tirpitz* had moved to Altenfjord. But the messages could not prove for sure that she was still there. The Admiralty, which knew from conventional sources that surface ships were primed to attack, decided the threat was too great. A fateful order was issued, that the convoy should scatter. Just beforehand, positive Enigma intelligence was delivered, showing that the *Tirpitz* was still at her berth.

PQ17 never made it intact to Archangel. Twenty-four merchant ships out of the 36 that sailed were sunk by U-boats and aircraft, which picked them off as defenceless singletons; 153 merchant seamen lost their lives.

Market Garden. After the Allies eventually broke out of Normandy in 1944, the collapse of the German army seemed imminent. Supreme Commander Eisenhower ordered advance across a broad front. On the left, Montgomery would drive towards the Ruhr, capturing bridges over the Maas and the Rhine. The plan would cut off the German Fifteenth Army in the Netherlands and also allow Monty to capture the V-2 launch sites in Zeeland. The operation would be launched on 17 September 1944.

But German generals von Rundstedt and Model knew some sort of attack was coming. Behind-the-lines rest and refit for the 9th and 10th SS Panzer Divisions was cancelled. On 15 September, a decrypt at Bletchley Park showed that Rundstedt and Model had moved their

Operation Market Garden: British paratroops of the 1st (British) Airborne Division are on their way to attempt the capture of the bridge over the Neder Rijn at Arnhem.

Field Marshal Erwin Rommel during the North African campaign.

```
40288   GROUP II/129
        TANGIER to MADRID
        RSS  148/28/9/42
        HKRO    7100k cs    1245 GMT    28/9/42
        268/5[7]5

[Nr. 268] Am 27. Um 20 Uhr Wache.  West=
einfahrt ein Kanonenboot, CARNERO ein Schnell=
boot.  2135, West-Ost dunkle Einheit.  2. 220
Ost-West, mittl. neutr. Frachter.  Am 28ten
um 0050, West-Ost,klein span. Frachter.  0115
West-Ost, k[l]ein span. Frachter.  0125 West-
Ost, ein kanonenboot.  0151 von West nach
TANGER, klein span. Frachter.  0.345, Ost-
West, mittl. neutr. Frachter.  0730, Ost-West,
span. Frachter.  08 Uhr Wache wie abends.
Maessige Sicht.
                        KRU[SE].
```

```
40394   GROUP II/129
        MADRID to TANGIER
        RSS 165/29/9/42
        ZEN on 6735 kcs.    1714 GMT. 29/9/42
        25/145 [ADCHE]

[Nr. 25] 29 September um 1350 Uhr 7 Geleitboote
aus BASTA ausgelaufen.  Wann und ob dort passiert?
                        SOMOZA.
```

```
40492   GROUP II/129
        TANGIER to MADRID
        RSS 13/30/9/42
        UPOO on 5300 kcs.    1913 GMT. 29/9/42
        272/115 [ADCVA]

[Nr. 272] Zu dort Spruch 25.  Von 13 bis 17
Uhr starker Nebel und Regen.  Absolut keine Sicht.
```

A dialogue enciphered by Enigma between the German spy in Tangier and his controller in Madrid, after the opening of the El Alamein campaign. Tangier sends reports of shipping movements through the Gibraltar straits. Madrid wants to know about escorts, but it was foggy and raining and nothing could be seen. © Crown Copyright – Used with permission of Director GCHQ.

headquarters to a new location only a couple of miles from the drop zones to be used by the British 1st Airborne Division.

Urgent messages sent to Ultra recipients were graded for priority: Z was the lowest priority and ZZZZZ the highest. Since the analysts at Bletchley were unaware of the forthcoming operation, the message went out as priority ZZ. The intelligence did not reach the commanders in the field, who remained unaware of the strength of the German opposition they would face. The attack failed, and only in March 1945 did Monty get across the Rhine.

All at sea – Naval Enigma

Naval Enigma was more difficult to solve than German army or air force Enigma. And, to the dismay of the codebreakers who were finding naval Enigma very tough, it became more important to Britain during the terrible years of the Battle of the Atlantic. Naval Enigma posed some special challenges:

- The German navy used eight, rather than five, rotors, which meant that there were 336 possible rotor orders (compared to 60 for the army and air force Enigma). The additional rotors had extra notches, giving more turnovers.
- Instead of relying on naval Enigma operators to come up with their own indicator- and message-settings, the German navy had a formal protocol. Before transmission the settings were disguised by adding two letters as padding and then encoded using a bigram table. The principle of the bigrams was to switch any given pair of letters into a different pair. This resulted in an indicator of two four-letter groups which had to be decoded in order to gain access to the actual message.
- In early 1942 U-boats began to use a 4-rotor version of the Enigma machine, including a special reflector, and making the 3-rotor Bombe obsolete almost overnight in the U-boat war.

British soldiers during the raid on Lofoten.

Operation Claymore – the Lofoten pinch

In March 1941 a small task force attacked Norwegian fish oil factories in the Lofoten Islands, which are situated about 100 miles north of the Arctic Circle. The object, as told to an eager public, was to get a much-needed victory, when Britain fought on alone against the mighty Germans. Lofoten had a factory which processed herring oil into glycerine which was then used to make munitions.

The weather was bad and the commandos were seasick on the way. But the raid was a success and the fish oil factories were blown up. However, the best prize of all was kept completely secret. The real reason for the raid had nothing to do with herring oil. The codebreakers had asked for the raid to grab the cryptological material from a German trawler called *Krebs* which was moored there. Although the commandos did not manage to seize the bigram tables, they got the daily keys for the month of February 1941, and this material enabled Hut 8 to reconstruct the bigram tables anyway. 'Suddenly,' says Professor Jack Copeland, the historian of Alan Turing's machines, 'Hut 8 was properly open for business.'

U-boat Enigma operators had to encipher the information on the starting positions of their rotors using bigram tables rather than the Enigma machine itself.

Tafel 25

AA = ZZ	BA = RC	CA = MF	DA = ZV	EA = OI	FA = SL	GA = CI	HA = IQ	IA = PK	JA = ZH	KA = SZ	LA = MY	MA = JJ
B = OM	B = CQ	B = DF	B = EI	B = CF	B = RM	B = EK	B = GC	B = DD	B = AW	B = JC	B = KV	B = AI
C = IH	C = AG	C = PS	C = YJ	C = FR	C = TN	C = HB	C = ZL	C = OB	C = KB	C = RT	C = ZD	C = TT
D = QS	D = LR	D = LP	D = IB	D = PQ	D = UK	D = AO	D = EM	D = CX	D = BK	D = IR	D = IO	D = BF
E = DN	E = ZX	E = YH	E = MH	E = BG	E = QM	E = ZN	E = JM	E = QG	E = YT	E = TD	E = NY	E = SA
F = VO	F = MD	F = EB	F = CB	F = ZT	F = VU	F = BT	F = CL	F = BW	F = CH	F = GM	F = HM	F = CA
G = BC	G = EE	G = FV	G = FT	G = AY	G = PO	G = IN	G = YP	G = NF	G = LV	G = QY	G = YV	G = UZ
H = TU	H = TS	H = JF	H = BZ	H = IJ	H = WP	H = DK	H = BM	H = AC	H = EL	H = FP	H = GO	H = DE
I = MB	I = UG	I = GA	I = HL	I = DB	I = OG	I = YN	I = KT	I = RA	I = XH	I = UU	I = OU	I = RY
J = YF	J = DL	J = BY	J = LN	J = LL	J = XQ	J = FQ	J = AU	J = EH	J = MA	J = DR	J = FM	J = ER
K = CS	K = JD	K = QQ	K = GH	K = GB	K = NW	K = JQ	K = XM	K = MS	K = FS	K = PG	K = XD	K = VH
L = KP	L = FX	L = HF	L = BJ	L = JH	L = YA	L = XO	L = DI	L = FW	L = WF	L = CM	L = EJ	L = FN
M = PW	M = HH	M = KL	M = NS	M = HD	M = LJ	M = KF	M = LF	M = SE	M = HE	M = VY	M = PE	M = QA
N = XY	N = OO	N = RG	N = AE	N = QW	N = ML	N = WM	N = FU	N = GG	N = OD	N = BQ	N = DJ	N = GR
O = GD	O = QU	O = XU	O = VM	O = YL	O = ZP	O = LH	O = WK	O = LD	O = EW	O = NA	O = WA	O = WN
P = RE	P = YC	P = NM	P = QO	P = SN	P = KH	P = VW	P = MQ	P = TH	P = ND	P = AL	P = CD	P = OZ
Q = FZ	Q = KN	Q = BB	Q = XS	Q = KR	Q = GJ	Q = FY	Q = VG	Q = HA	Q = GK	Q = OS	Q = VB	Q = HP
R = NQ	R = VQ	R = UE	R = KJ	R = MJ	R = EC	R = MN	R = NI	R = KD	R = VC	R = EQ	R = BD	R = PC
S = SX	S = XW	S = AK	S = CZ	S = WZ	S = JK	S = UM	S = UO	S = UQ	S = IT	S = MV	S = QZ	S = IK
T = WV	T = GF	T = SU	T = RI	T = RK	T = DG	T = NK	T = OQ	T = JS	T = PI	T = HI	T = AZ	T = XB
U = HJ	U = NO	U = WX	U = SP	U = TY	U = HN	U = TL	U = TJ	U = VE	U = TF	U = WC	U = TA	U = NB
V = UI	V = WT	V = OK	V = WR	V = UA	V = CG	V = OI	V = ZJ	V = US	V = LB	V = VG	V = JG	V = KS
W = JB	W = IF	W = TP	W = PY	W = JO	W = IL	W = SG	W = YR	W = QE	W = XF	W = WH	W = UX	W = ZO
X = SR	X = PU	X = ID	X = UC	X = NU	X = BL	X = PM	X = RO	X = WH	X = SC	X = YI	X = KZ	X = ZO
Y = BG	Y = CJ	Y = VS	Y = ZR	Y = OX	Y = GQ	Y = QK	Y = GZ	Y = HZ	Y = QB	Y = ZF	Y = SV	Y = LA
Z = LT	Z = DH	Z = DS	Z = TW	Z = VJ	Z = AQ	Z = HY				Z = PW		Z = NX

The German navy introduced a 4-rotor version of the Enigma machine in early 1942.

Breaking 4-rotor Enigma keys was ultimately too difficult using 3-rotor Bombes. This is an example of a 4-rotor fast Bombe designed by Doc Keen.

Four-rotor Bombes were also designed separately in the United States. This one survived the war and is on display at the National Cryptologic Museum there.

The Bletchley Park codebreakers faced up to this formidable set of problems in a variety of ways. Alan Turing discovered how the bigram system worked during late 1939 by analysing 100 old messages. But he didn't actually have the bigram tables, so the Royal Navy set up a series of 'pinches' to recover these – or other materials which enabled them to be reconstructed – from small Enigma-carrying craft such as weather ships.

When the 4-rotor Enigma was introduced, the codebreakers tried a variety of ways to adapt 3-rotor Bombes to cope with the fourth rotor. But efficient codebreaking of 4-rotor Enigma required the development of a special 4-rotor Bombe, with a super-fast processing speed. Doc Keen adapted his Bombe to produce a super-fast version, over 60 of which were produced to assist Bletchley Park. Other 4-rotor Bombes were separately designed and built in the USA, and a bank of 4-rotor Bombes operated in Washington DC until the end of the European war.

In Hut 8 the codebreakers battled with the U-boat version of Enigma. Alan Turing's room was here during the Battle of the Atlantic; he is reported to have chained his tea-mug to the radiator to prevent it being taken away.

Rescue and rebuild

The war in Europe was over in May 1945, but the secret had to go on. The first step in preserving the secret was to dismantle the machines which, by their very existence, told a tale. The Bombes were ordered to be broken up.

Bletchley Park itself continued for a while to be a training school for the Government Code & Cypher School – now renamed GCHQ – but over time the site was used variously as a training centre for the GPO (which later became British Telecom), a teacher training college, and offices for the Civil Aviation Authority. The huts and blocks fell into disrepair; some hideous interior décor was added to the Mansion, which became tatty and dirty. By the early 1990s Bletchley Park was unkempt, unloved, and due for demolition. But a major effort by local people keen to preserve the heritage of the codebreakers eventually saved the

Veteran Bombe operator Jean Valentine explains its workings to Her Majesty the Queen in 2011.

A British Tabulating Machine Co. blueprint from the Bletchley Park Archive. © Crown Copyright – Used with permission of Director GCHQ.

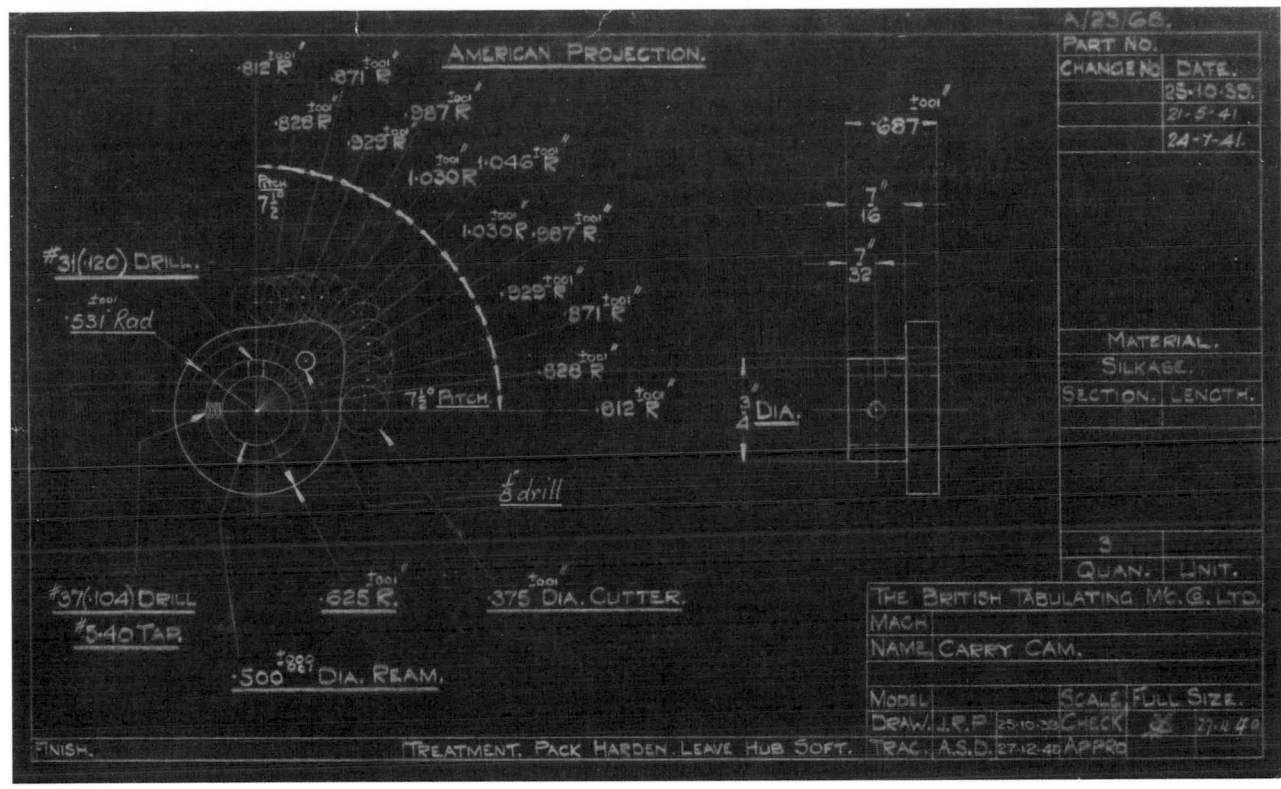

site. Since 2013 a generous investment of the Heritage Lottery Fund has enabled substantial restoration work to proceed.

Meanwhile, a Bombe was being painstakingly reconstructed. A team supported by the Computer Conservation Society and led by John Harper began work in 1995. GCHQ returned a collection of technical drawings relating to the Bombe to Bletchley Park, where they now constitute part of the Archive. By 2006 a replica Bombe was ready to be commissioned and HRH the Duke of Kent officially switched on the new Bombe on 17 July 2007, 62 years after the Bombes at Bletchley Park ceased to help break German Enigma codes.

Bombe disposal

'In 1945 the German War was over and suddenly we were redundant. Watchkeeping ceased and we were set to work dismantling the Bombes. I recall sitting at trestle tables unscrewing the drums.'

Jenny Conduit, WRNS
quoted by Gwendoline Page, ed., *We Kept the Secret*

A reunion of Bletchley Park veterans photographed outside the Mansion in 2013.